THE HISTORY OF VIDEO GAMES

CHARLIE FISH

WHITE OWL
AN IMPRINT OF PEN & SWORD BOOKS LTD.
YORKSHIRE – PHILADELPHIA

Dedicated to Emma, for making everything possible.

We bonded over Scrabble and *Super Mario 64*.

First published in Great Britain in 2021 by
White Owl
An imprint of
Pen & Sword Books Ltd
Yorkshire - Philadelphia

Copyright © Charlie Fish, 2021

ISBN 978 1 52677 897 0

The right of Charlie Fish to be identified as author of this work has been asserted by him in accordance with the Copyright, Designs and Patents Act 1988.

A CIP catalogue record for this book is available from the British Library.

All rights reserved. No part of this book may be reproduced or transmitted in any form or by any means, electronic or mechanical including photocopying, recording or by any information storage and retrieval system, without permission from the Publisher in writing.

Typeset in 10.5/12.5 pts Adobe Devanagari
by SJmagic DESIGN SERVICES India.

Printed and bound by Printworks Global Ltd, London/Hong Kong

Pen & Sword Books Ltd. incorporates the Imprints of Pen & Sword Books: Archaeology, Atlas, Aviation, Battleground, Discovery, Family History, History, Maritime, Military, Naval, Politics, Railways, Select, Transport, True Crime, Fiction, Frontline Books, Leo Cooper, Praetorian Press, Seaforth Publishing, Wharncliffe and White Owl.

For a complete list of Pen & Sword titles please contact

PEN & SWORD BOOKS LIMITED
47 Church Street, Barnsley, South Yorkshire, S70 2AS, England
E-mail: enquiries@pen-and-sword.co.uk
Website: www.pen-and-sword.co.uk

or

PEN AND SWORD BOOKS
1950 Lawrence Rd, Havertown, PA 19083, USA
E-mail: Uspen-and-sword@casematepublishers.com
Website: www.penandswordbooks.com

CONTENTS

Level 1   Introduction 4

Level 2   Platforms and Technology 6

Level 3   People and Personalities 39

Level 4   Companies and Capitalism 62

Level 5   Gender and Representation 73

Level 6   Culture and Community 83

Level 7   Games 100

          References 115

          Index 117

Level 1
INTRODUCTION

While researching this book I've been blown away, again and again, by the scale of the video game industry. But, despite the immensity of this commercial juggernaut, one inspired individual with a quirky vision is so often able to create something new that captures the hearts of millions.

I was born in 1980, so video games have always been part of my life. I was the perfect age to be swept up by Nintendo's ascendancy. When I was seven years old, living in Massachusetts, I couldn't wait for school to finish so I could go to Danny's house to play on his Nintendo Entertainment System.

At home, we had an IBM personal computer. Our floppy disks were still floppy. I bonded with my dad over Sierra On-Line's graphic adventure games. Later, having moved to the UK, I filled notebooks with level designs for *Lemmings* and *Commander Keen*.

At the age of 13, my parents rewarded me for good exam results by buying me a Super Nintendo Entertainment

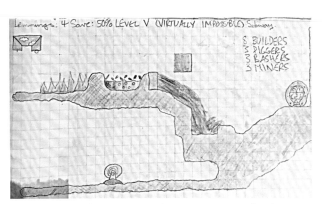

A level design for *Lemmings* from my exercise book, 1991. (Charlie Fish / CC BY 4.0)

System with *Super Mario World* – I could hardly believe the enormity of this gift. Meanwhile, on the PC, I was building urban utopias in *SimCity 2000* and (to my parents' distaste) blasting hell demons in *Doom*.

SimCity 2000 (1993). (Maxis)

As I grew older, I made a conscious effort to spend less time playing video games, but never managed to resist their siren song for long. Every few years I bought the latest Nintendo console, and embraced the joy of gaming again for a few intense months. The advent of smartphones ensnared me into the habit of playing a quick game whenever I had a few spare minutes – and too often when I didn't.

While writing this book, I was designing levels in *Super Mario Maker 2*, building civilizations in *Through the Ages*, and pretending to be a moody goose in *Untitled Goose Game*. Games are rewarding on a deep-seated psychological level. Just as rewarding are the stories behind them. I hope you'll find these stories as fascinating as I do.

Untitled Goose Game (2019). *(House House/Panic)*

My dad's first computer, a Sol-20 he bought in Toronto, Ontario in 1976. *(Robin Sundt)*

Level 2
PLATFORMS AND TECHNOLOGY

Video games are shaped by the technology of the day, and human creativity has always found ingenious ways of pushing the limitations of the hardware.

In 1990, British student Andy Davidson was playing around with a Casio graph-plotting calculator during a particularly boring maths lesson, when his friend said, 'I bet you can't make a game on that.' He accepted the challenge. He made a basic artillery game, and soon ported it over to the school's Amiga computer. The game was so popular, his teacher banned it. After working on it for four more years, he approached struggling game developers Team17, who snapped up the idea, and the hugely successful *Worms* franchise was born.

The idea behind *Worms* was not new. Artillery games, usually featuring two players taking turns to fire projectiles at each other on the screen, have a long heritage – a watershed example being *Artillery Simulator* released in 1980 for the Apple II home computer.

In fact, the world's first video game was almost an artillery game. In 1947, American television pioneer Thomas T. Goldsmith, Jr. and Estle Ray Mann filed a patent for a 'Cathode-ray tube amusement device'. The idea was to hook up an oscilloscope to a TV screen so the user could control the cathode-ray tube's electron gun with dials, like an Etch-a-Sketch, and simulate firing a missile at a target. But the equipment was expensive and the device never hit the market.

There were a handful of early computers designed to play games, using a series of light bulbs for the display. First was the *Nimatron*, presented at the New York World's Fair in

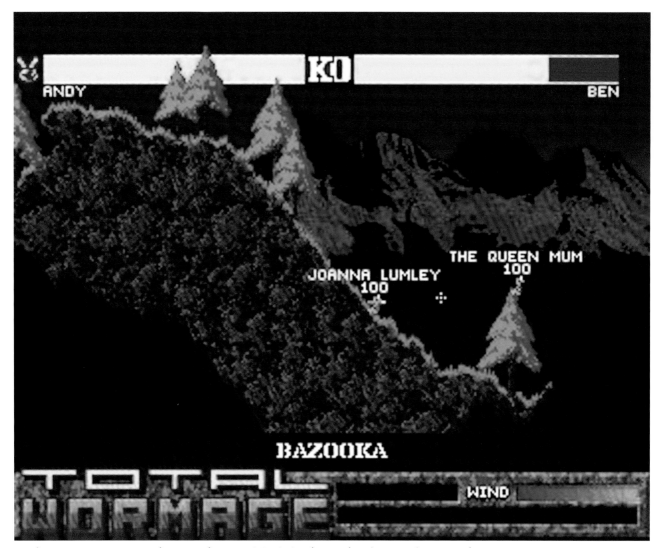

Total Wormage, precursor to the original *Worms* (1995). *(Andy Davidson/Team 17/Ocean Software)*

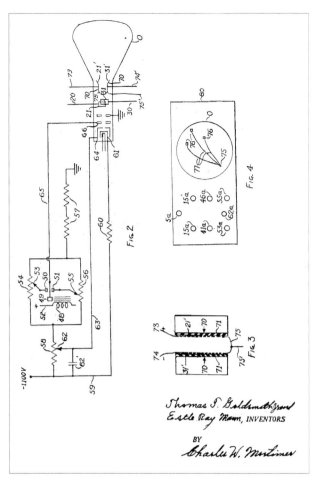

Patent for 'Cathode-ray tube amusement device', 25 January 1947. (Thomas Tolivan Goldsmith Jr. – US Patent 2455992, Public Domain)

The designer of the *Nimatron*, celebrated quantum physicist Edward Condon, considered it a 'shameful' failure that he did not exploit the full potential of his idea. (Manuscripts and Archives Division, The New York Public Library. 'Westinghouse – Woman with electric brain machine' The New York Public Library Digital Collections. 1935 – 1945)

April 1940, which played the mathematical strategy game *Nim*. The *Nimatron* was a success – about 50,000 people played it, and the computer won more than 90 per cent of the games. But in subsequent years it was largely forgotten, and a decade passed before another game-playing computer made a splash.

At the Canadian National Exhibition in 1950, *Bertie the Brain* was a four-metre-tall monstrosity that emitted a loud buzz, and featured a visual display backlit with light bulbs. The machine was purpose-built to play noughts and crosses (a.k.a. tic-tac-toe), with several difficulty levels. It was very popular at the show and, by all accounts, hard to beat.

A similarly gargantuan machine was built in Britain a year later, also relying on circuitry made of vacuum tube valves, inspired by the *Nimatron*. The display still relied on light bulbs. Not until 1952 were the first software-based video games created that displayed visuals on an electronic screen – a draughts program and a noughts and crosses program, both made independently by British computer scientists.

Arguably, these programs don't count as video games, because they lacked

American comedian Danny Kaye beats *Bertie the Brain*, after the difficulty was turned down a few notches (1950). (Bernard Hoffman/The LIFE Picture Collection/Getty Images)

Pool (1954) on the MIDSAC computer. ('Willow Run Research Center, MIDAC Computer, 1954; BL018366.' University of Michigan Library Digital Collections)

moving graphics. The first game with graphics that updated in real-time, rather than only when the player made a move, was a pool simulation created on the University of Michigan's room-sized 'Michigan Digital Special Automatic Computer' (MIDSAC) in 1954. The player used a joystick to line up a cue on a 33-centimetre 'fluorescent screen', then the computer made 25,000 calculations a second to show the balls moving around the table, bouncing off each other and disappearing into the pockets. (For comparison, the smartphone in your pocket can manage over 3 billion calculations per second.)

Following in the footsteps of these pioneers, games continued to be used to showcase new technology and computing power. The first video game created purely

Tennis for Two (1958) on a DuMont Lab Oscilloscope Type 304-A. *(Brookhaven National Laboratory (BNL) / Public domain)*

for entertainment value was American physicist William Higinbotham's *Tennis for Two*, created in 1958 for an exhibition at Brookhaven National Laboratory. The game ran on an analogue computer, about the size of a microwave oven. The oscilloscope display showed the tennis court as a horizontal line, with a short vertical line sticking up in the middle representing the net. Players used custom-made aluminium controllers, and the game simulated air resistance and gravity.

The rise of electronic transistors and magnetic-core memory meant computers were gradually getting smaller and more affordable. In late 1959, the Digital Equipment Corporation released the PDP-1 computer, costing a mere $120,000 (the equivalent of around $1,000,000 in 2020 dollars) and weighing in at 730 kilograms. When the Massachusetts Institute of Technology acquired a PDP-1, the playful members of MIT's Tech Model Railroad Club decided to create a game on it, inspired by pulp science fiction writers such as E. E. 'Doc' Smith. They called it *Spacewar!*

Throughout the 1960s, students and researchers played *Spacewar!* and ported it to other systems as

Spacewar! (1962) running on the California Computer History Museum's PDP-1. *(Joi Ito from Inbamura, Japan / CC BY)*

'minicomputers' became cheaper and more widely available. Two players simultaneously controlled spaceships, trying to blast each other while avoiding the gravity well of a small sun. *Spacewar!* was the first truly influential video game – it directly inspired the first two coin-operated arcade video games, *Computer Space*

Creator of *Spacewar!* Steve Russell with a PDP-1. *(Alex Handy [cropped by Arnold Reinhold] / CC BY-SA)*

and *Galaxy Game* in 1971, which in turn inspired a whole industry of arcade video games.

Up to the 1970s, amusement arcades were busy places where young people with pocketfuls of old pennies or dimes could load up a jukebox, consult Zoltar the fortune teller, practise at a shooting gallery, or play an electronic pinball machine. Video games, starting with the wildly successful *Pong* in 1972, helped arcades shed their iniquitous associations with gambling and organised crime, and rapidly dominated the arcade scene.

Amusement arcade operators embraced the new trend. Video game cabinets had fewer moving parts, so broke down less often, than the electro-mechanical games they replaced; and made way more money. By September 1974 there were an estimated 100,000 coin-operated video games in the US, making a quarter of a billion dollars of revenue a year (that's about $1.3bn in 2020 dollars).

These arcade video games relied on dedicated hardware rather than computers. *Pong*, in which two players each controlled a paddle trying to keep the digital ball on-screen, was hugely influential. As more companies started competing in the market, technological innovation followed. In 1975, *Gun Fight* was the first game to use a microprocessor, Intel's 8-bit 8080. Two players controlled cowboys trying to shoot each other across a field of cacti. The Japanese designer of *Western Gun*, which was the name of *Gun Fight* before it was licensed in the US, was impressed by the improved graphics and smoother animation the microprocessor afforded. He used the same technology to

Pong (1972) cabinet, signed by designer Allan Alcorn. *(Chris Rand Derivative work: Georgfotoart / CC BY-SA)*

create the game that heralded arcade video gaming's golden age, 1978's *Space Invaders*.

Meanwhile, home video gaming had a parallel evolution. In 1950, 9 per cent of homes in the US had a television set – by 1960, almost 90 per cent. The UK lagged a little further behind, with TVs in just over two-thirds of homes in 1960.

German-born American engineer Ralph Baer had the idea to make an interactive game on a TV as far back as 1951, but the company he worked for had no interest in allowing him to pursue the idea. Many years later, as colour televisions were becoming popular and electronics cheaper, he tried again. In August 1966 he wrote a four-page proposal

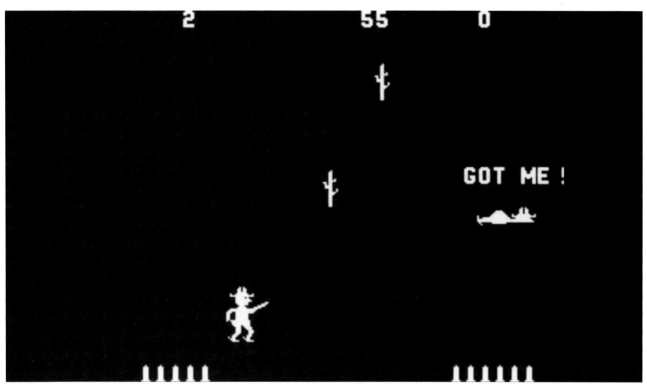

Gun Fight (1975). *(Taito/Midway)*

Space Invaders (1978), pictured in YESTERcades arcade in Somerville, New Jersey, alongside later games. *(Rob DiCaterino / CC BY 2.0)*

Ralph Baer's 'Brown Box' prototype. *(Philip Steffan / CC BY-SA)*

for a TV 'game box', but rather than give it to his bosses at Sanders Associates (a defence contractor, now part of BAE Systems) he assigned one of his technicians to develop the idea, and over the next four months they created a prototype.

The company's director of research and development saw enough potential in the prototype to fund the project for $2,500. Baer and his colleagues set to work. Several iterations later, the product was licensed to Magnavox, a company that manufactured televisions, radios and phonographs.

It took until 1972 for Magnavox to release the Odyssey, the world's first home gaming console. It hit the market two months before *Pong* swept the nation and then the world. The Magnavox Odyssey retailed at $100 ($620 in 2020 dollars), much more expensive than the $20 Baer had originally envisaged, and even to get to that price there were compromises: the console's colour circuitry had been discarded and replaced by cheap plastic overlays that could be attached to the TV screen. The console came with 13 built-in games (out of a total of 28 ever made for the platform), two controllers with rotary dials and switches, and a light gun accessory for target-shooting games – but also paper money, playing cards and poker chips to supplement the primitive gameplay.

The idea of being able to control an object on the screen of your own TV was incredible at the time, and the console sold well, shipping about 350,000 units worldwide before it was discontinued in 1975 – by which time other companies had their eye on the home gaming market.

The first generation of home video game consoles – all packaged with a variant of *Pong* – comprised simple systems that jacked into your television via coaxial cable, and some handheld devices, that were restricted to their built-in games. Most of the first generation were released in 1975–78, such as the UK's Binatone TV Master which launched for £39.95 (£250 in 2020 pounds) and thrilled children with its bright yellow packing and lurid orange plastic console.

Magnavox Odyssey (1972). *(Evan-Amos / Public domain)*

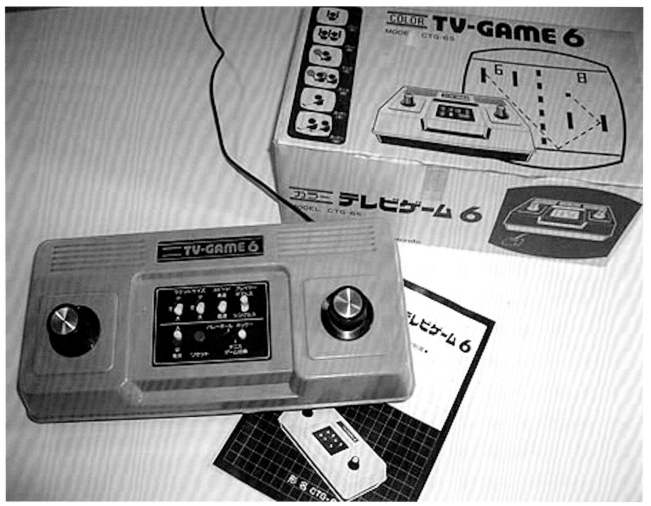

Nintendo's first console, the Color TV-Game 6 (1977). Nintendo is the only manufacturer from the first generation of video game consoles that is still making consoles today. *(Vinelodge / CC BY-SA)*

The best-selling first generation consoles were Coleco's Telstar series, which shipped a million units, and Nintendo's five Color TV-Game units, which sold an impressive 3 million exclusively inside Japan. More than 200 companies jumped on the bandwagon, over half of which only ever released one video game console. Of those 200, only about 20 companies stuck around for the second generation of video game consoles – the rest never entered the market again.

Second generation consoles were distinguished by the games being available to buy separately on programmable Read-Only Memory (ROM) cartridges, rather than being built into the machine. This innovation started with the Fairchild Video Entertainment System, later renamed the Fairchild Channel F (the F stood for Fun), in 1976. The Fairchild Channel F was the first console to use a microprocessor – another feature of second generation consoles.

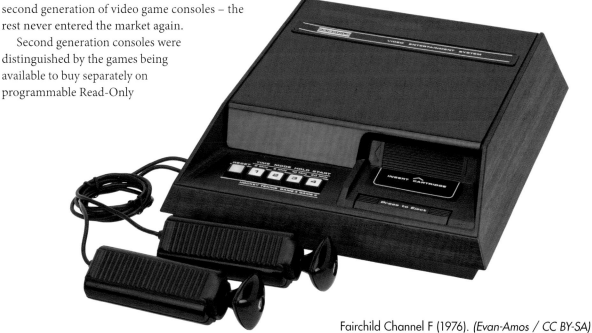

Fairchild Channel F (1976). *(Evan-Amos / CC BY-SA)*

Jerry Lawson (centre) designed the Fairchild Channel F video game console and pioneered the commercial video game cartridge. Pictured with radio host Drew Verbis. *(OU133 / CC BY-SA)*

The best-selling second generation consoles were the ColecoVision, the Magnavox Odyssey² (each sold 2 million), the Intellivision from Mattel Electronics (3 million), and the runaway leader, the Atari Video Computer System (VCS), later renamed the Atari 2600, which stayed on the market until 1992, selling a total of 30 million units.

Being able to buy a console once and then buy additional games for it relatively cheaply was so appealing that the bottom dropped out of the market for first generation consoles that were restricted to their built-in games, and huge amounts of remaining stock were sold off at a loss. This triggered a crash in 1977–78 which wiped out many of the second generation wannabes, including the precocious Fairchild Channel F, which was discontinued after struggling to sell 250,000 units.

To begin with, sales of the Atari VCS also suffered, but all that changed

The European version of the Magnavox Odyssey² (1978) was called the Philips Videopac G7000. *(boffy_b / CC BY-SA)*

Mattel's Intellivision (1979), with games. *(Bidou; NaSH / CC BY-SA)*

when they started porting successful arcade games. *Space Invaders*, licensed from Taito and released on the Atari VCS in 1980, was a 'killer app', helping to quadruple sales of the VCS over the next two years. In this format, *Space Invaders* was the first game to sell over a million copies, eventually doubling that total. The VCS versions of Atari's own arcade games *Missile Command* and *Asteroids*, released in 1981, sold about 2.5 million and 3.8 million units respectively. But the most successful arcade game adaptation (in terms of sales, if not gameplay) was *Pac-Man*, licensed from Namco and released for the Atari VCS in 1982 – it sold over 7 million copies.

An Atari VCS running *Jr. Pac-Man* (1983). *(Georges Seguin [Okki] / CC BY-SA)*

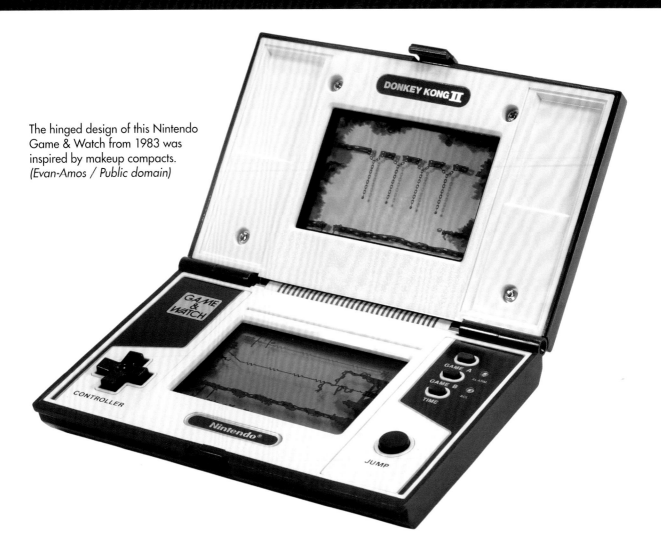

The hinged design of this Nintendo Game & Watch from 1983 was inspired by makeup compacts. *(Evan-Amos / Public domain)*

Meanwhile, video gaming was tentatively blossoming in other formats. The handheld gaming market saw the introduction of Nintendo's enduring Game & Watch series, each featuring a single game played on an LCD screen (ultimately 60 games were released, which collectively sold over 43 million units).

And home computers were becoming more popular, featuring video games alongside productivity software and the ability for users to create their own programs. Back then, 5¼ inch computer floppy disks were capable of storing 1.2 megabytes (these days a USB drive the size of your thumbnail can store 400,000 times as much). By 1982, there was a home computer in over 600,000 American households, and they were retailing at around $530 ($1,400 in 2020 dollars). Compare the Atari VCS, which launched at $199 and was selling for $125 by 1982.

Floppy disks: 8 inch disk (1971), 5¼ inch (1976) and 3½ inch (1986). *(George Chernilevsky / Public domain)*

Like *Star Wars*, Pac-Man had its own merchandising empire. (~ tOkKa / CC BY-SA 2.0)

Author's sister-in-law playing *Galaga* (1981). *(David Sundt)*

But in the US, arcades rather than homes were at the forefront of the video game revolution. The scale of the American obsession with arcade video games during their golden age cannot be understated. The *Pac-Man* arcade game grossed over $1 billion in the US ($3bn in 2020 dollars) in its first year of release alone – all in quarters. By 1982, an estimated 7 billion coins had been inserted into 400,000 *Pac-Man* arcade machines worldwide. For the sake of comparison, that's considerably more than the worldwide box office takings of *Star Wars* and *The Empire Strikes Back*, which were the two highest grossing films ever at the time.

It was not to last. The American video gaming market was about to experience a colossal recession. In 1983, US video game industry revenues were around $3.2 billion. By 1985 they had dropped 96 per cent to around $100 million. Globally, the video game market dropped from $42 billion in 1982 to $14 billion by 1985.

The causes of the dramatic crash were threefold. First, the market was flooded – arcade owners were buying more machines than their patrons could support, and stores were overflowing with knock-off consoles. Second, console manufacturers could no longer rely on releasing their own new games as an income stream because third parties were making cheap unlicensed variants. Third, home computers were on the rise, led by the Commodore 64 in the US and the Sinclair ZX Spectrum in the UK, stealing video game market share.

Commodore 64 (1982). *(PrixeH / CC BY-SA)*

The crash had a lesser effect in the UK, where home computers had been adopted more quickly than consoles – in 1983 the UK boasted the highest level of computer ownership in the world. And Japan's video game market escaped the crash unscathed, allowing them to take over dominance of the global market in subsequent years.

Three 8-bit video game consoles were released in Japan in 1983, kicking off the third generation: Nintendo's Family Computer (Famicom), Sega's SG-1000, and Microsoft Japan's MSX hybrid computer-console. They proved hugely popular, buoyed by a bullish economy in Japan and unchallenged by the previously ubiquitous American manufacturers. By 1986, one in five Japanese homes had a Famicom.

At the Consumer Electronics Show in 1985, Nintendo unveiled the American version of its Famicom, called the Nintendo Entertainment System (NES). American distributors were so burned by the video game crash that Nintendo went to great lengths to package their product as anything but a console. Game cartridges were front-loaded into the machine, intended to be reminiscent of video cassette recorders, which were growing in popularity. Early marketing leaned heavily on a short-lived accessory called R.O.B. (Robotic Operating Buddy), an E.T.-like robot that reacted to flashes on the TV screen.

The marketing worked. After the NES saw a wide release in 1986, it flew off the shelves, and helped to revive the American video game industry. In 1988 alone, Nintendo

Nintendo's Family Computer (1983), which when exported became the Nintendo Entertainment System. *(Evan-Amos / Public domain)*

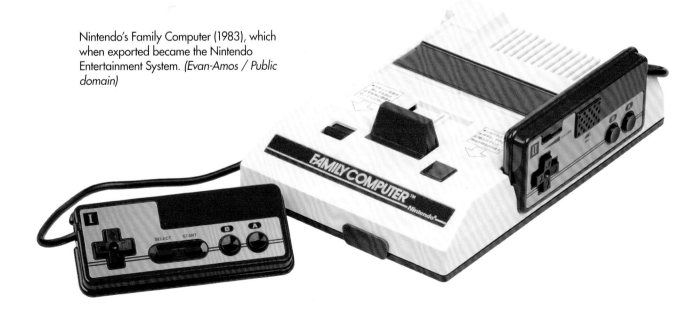

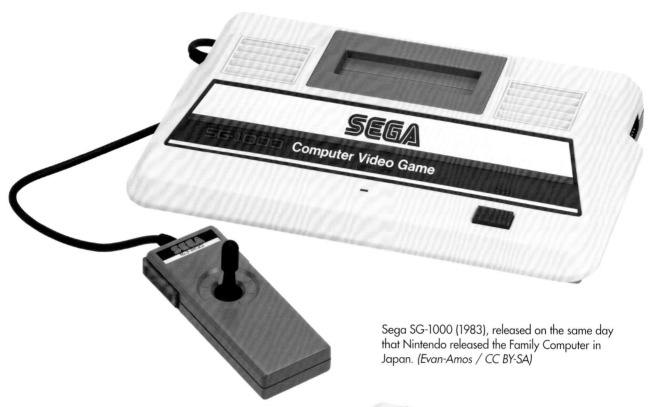

Sega SG-1000 (1983), released on the same day that Nintendo released the Family Computer in Japan. *(Evan-Amos / CC BY-SA)*

Nintendo's R.O.B. *(Evan-Amos / Public domain)*

Nintendo Entertainment System (1985). *(Evan-Amos / Public domain)*

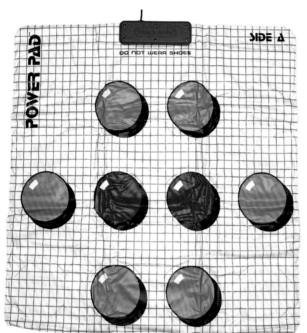

Nintendo released several accessories for the NES, such as the Family Trainer (1986), released in the US as the Power Pad in 1988... *(Evan-Amos / Public domain)*

sold 7 million NES systems, almost as many as the popular Commodore 64 personal computer sold in its first five years. In 1989 the NES overtook the Atari 2600 (formerly called Atari VCS) as the best-selling video game platform of all time. By 1990, 30 per cent of American households owned a NES, compared to 23 per cent owning a personal computer.

Nintendo had learned lessons from the American video game crash and kept a tight rein on third party game developers, preventing them from releasing any games developed for the NES onto other consoles. In an effort to enforce a strict licensing system that would become industry standard, each NES console and game cartridge included a region-specific lockout chip which had to match for the game to work. Nintendo manufactured all cartridges themselves, with publishers being required to pay for them in full (non-refundable) before they were manufactured.

Nintendo's main competitors in the 1980s were Sega, whose Master System had superior hardware and dominated in the European and South American markets, and NEC Home Electronic's TurboGrafx-16 (known in Japan and France as PC Engine) which gained particular traction in Japan. This heated three-way rivalry would

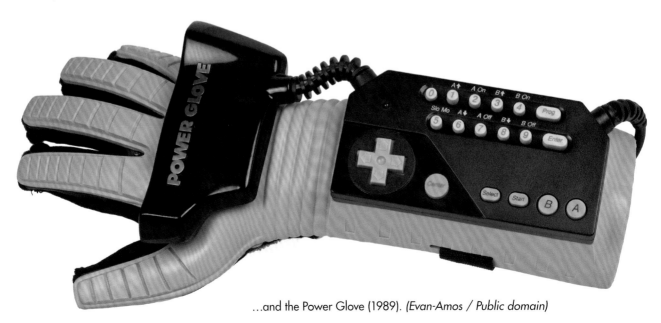

...and the Power Glove (1989). *(Evan-Amos / Public domain)*

NES game cartridges. *(Tony Webster from Portland, Oregon, United States / CC BY)*

continue through the 1990s, into the fourth generation of video game consoles – although the Turbografx-16 would end up in a distant third place.

The battle between Nintendo and Sega reached fever pitch during the fourth generation of consoles, with the release of the Sega Mega Drive (known in the US as Genesis) and the Super Nintendo Entertainment System (SNES). In the US, where Nintendo had 90 per cent market share at the end of the 1980s, Sega fought hard to position itself as the punk rebel to Nintendo's boring parent, with huge success. Sega's new mascot, the spiky and speedy Sonic the Hedgehog, was prickling with attitude – the polar opposite of Nintendo's moustachioed Italian plumber Mario. There were many playground confrontations over which was better, Nintendo or Sega.

Sega adopted an aggressive marketing strategy, turning launches of new games into flashy media events. And Sega found a niche in the market by offering much greater freedom to game developers and encouraging more adult content, both in stark contrast to Nintendo. When both consoles released the hugely popular fighting game *Mortal Kombat* in 1993, Nintendo's sanitised version with grey blood was outsold by Sega's version with red blood, allowing Sega to eclipse Nintendo as market leader in the US for the first time. The Sega Mega Drive also did well in Europe, but less so in the Asian markets.

Midway's *Mortal Kombat* was notable for several reasons. Along with Capcom's *Street Fighter II*, it helped inject new life into the beleaguered arcade gaming industry at the start of the 1990s. And, when released for home consoles, the depiction

Japanese Sega Mega Drive (1988) *(Evan-Amos / Public domain)*

European Super Nintendo Entertainment System (1992, although the original Japanese version was released in 1990) *(Sandos / CC BY-SA 3.0)*

Sonic the Hedgehog (1991) on the Sega Mega Drive. *(BagoGames / CC BY 2.0)*

Super Mario Kart (1992) on the SNES. *(Nintendo)*

'Sega does what Nintendon't'. *(Sega of America)*

of violence in *Mortal Kombat* caused sufficient controversy in the US that the Entertainment Software Rating Board was created to give film-style age ratings to video games.

Nintendo and Sega were also battling in the handheld gaming arena, but that fight was a whitewash. Sega's Game Gear was launched internationally in 1991 and sold 11 million units globally until it was discontinued a decade later – obliterating NEC's TurboExpress (1.5 million sold) and Atari's Lynx (0.5 million sold). But Nintendo's Game Boy, launched in 1989, left them all in the shade, shifting a total of 119 million units (including Game Boy Color, launched in 1998). Despite not having a colour screen, as offered by its competitors, the Game Boy won out thanks to greater battery life, cheaper price tag, smaller size, and wider third party support. Plus, of course, its ace in the hole – it was packaged with what became the best selling game of all time: *Tetris*.

The runaway success of home consoles meant computer platforms were not a priority for game developers. IBM PCs, formerly aimed at businesses only, had by 1987 overtaken the Apple II and the Commodore 64 as a video game platform. In 1989 the VGA card was released, allowing for higher resolution 640 x 480 graphics (more than twice that of home consoles at the time). The adoption of VGA cards was driven by the success of games such as Sierra's *King's Quest* series, in which the player solved puzzles by navigating around a fairy-tale landscape.

The original Game Boy (1989), on the top left, has been re-released in many varieties. *(JCD1981NL / CC BY 3.0)*

VGA graphic adventure games *Space Quest: Chapter I – The Sarien Encounter* (1986), *Loom* (1990) and *King's Quest IV: The Perils of Rosella* (1988). *(Digital Game Museum / CC BY 2.0)*

In 1991 four guys in Dallas, Texas founded id Software, innovating on the PC platform to make more console-like video games, with such seminal titles as *Commander Keen*, *Wolfenstein 3D* and *Doom* (read more about those games later, in Level 3: People and Personalities). But PC gaming was still overshadowed by consoles – in 1993, in the US, computer games made $430 million in revenue ($760m in 2020 dollars), versus $6 billion for console games. And arcade games made more than both of those put together. Compare the film industry, which in the US in 1993 generated $5 billion in revenue.

The introduction of the fifth generation of video game consoles, which were launched between 1993–1996, didn't do much to slow down the sales of the most popular fourth generation consoles. Even the third-generation NES was still selling well until it was discontinued in 1995. The battle between Sega and Nintendo raged on with the release of the Sega Saturn and the Nintendo 64. But the real story was the nascence of a major new player in the market: Sony.

At the Consumer Electronics Show in Chicago in 1991, Sony proudly announced that it was working with Nintendo to create a CD-ROM add-on for the Super

Resident Evil (1996) on the Sony PlayStation. *(Capcom)*

Crash Bandicoot (1996) on the Sony PlayStation. *(Naughty Dog / Sony Computer Entertainment)*

NES, provisionally called the Nintendo PlayStation. The companies had been working together in secret since 1989 on the project. The very next day, Nintendo got on stage and announced that they were dropping Sony in favour of their Dutch rival Philips. The SNES CD-ROM peripheral never made it to market.

Sony was furious about Nintendo's very public snub, and promptly took their technology to Sega. Tom Kalinske, Sega of America's CEO, explained the proposition: 'We had the Sony guys and our engineers in the United States come up with specs for what this next optical-based hardware system would be, and we met with Ken Kutaragi [the CEO of Sony's new video game division]. He said it was a great idea, and as we all lose money on hardware, let's jointly market a single system – the Sega/Sony hardware system.'

But Sega of America's true rival wasn't only Nintendo; it was parent company Sega of Japan. Sega of Japan occasionally overruled Tom Kalinske's big ideas and, foolishly, this was one of those times. Worse for Sega, their own CD-ROM accessory for the Sega Mega Drive console ended up being an expensive failure. Having been rejected by Sega and Nintendo, Sony decided to go it alone and develop their own console.

The first CD-ROM consoles capable of 3D graphics arrived in 1993. Panasonic launched the prohibitively expensive 3DO Interactive Multiplayer (launched at $699, which is $1,250 in 2020 dollars). Atari launched the Jaguar, which flopped due to hardware bugs, poor developer support tools and complicated

A model 2 Sega CD with a model 2 Genesis and a 32X attached. Each device required its own power supply. *(Evan-Amos / Public domain)*

multi-chip architecture that made the platform difficult to develop for – prompting Atari to leave the console market for good. Outside of the US, the Amiga CD^{32}, developed by Commodore, was briefly popular in the UK during the 1993 Christmas season, but barely lasted six months before it was discontinued as Commodore went bankrupt.

In May 1995, at the first Electronic Entertainment Expo (E3), there was a huge amount of buzz around the Sony PlayStation, which was due to be released in the North American market that autumn. Sega of Japan panicked, and decided to bring forward the launch of the Sega Saturn so they could get the jump on Sony. Sega of America CEO Tom Kalinske took to the stage and surprised everybody by announcing that the Saturn, also due to be released in the autumn, was already on the shelves, retailing at $399.

But Sega's gambit backfired. Sony's head of development Steve Race responded by walking up to the stage and saying just three words: the release price of the upcoming PlayStation. 'Two nine nine.' $100 cheaper.

Rayman (1995), a launch title for the Sony PlayStation's release outside of Japan. *(Ubi Studios / Ubi Soft)*

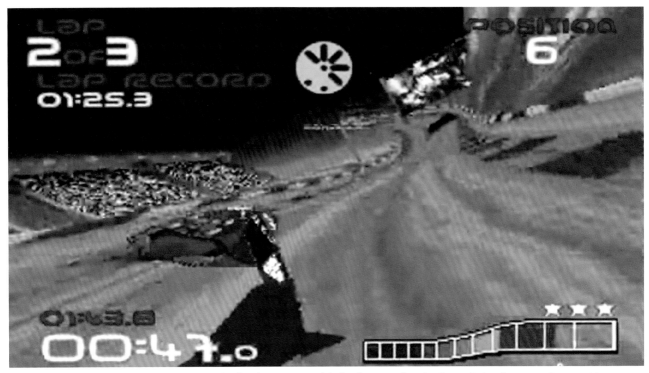
The enigmatic racing game Wipeout (1995) (stylised as wipE'out") helped secure the PlayStation's 'cool' image. *(Psygnosis)*

That, along with supply problems because the Saturn production line wasn't ready for its premature release, helped doom the Saturn to mediocre success, and ended Sega's ascendancy. The Nintendo 64, despite being cartridge based rather than having an optical drive, ended up outselling the Saturn almost three to one, and the PlayStation ended up being the most successful non-handheld of all time – until it was beaten by the PlayStation 2.

In the handheld market, there were no serious contenders for the Game Boy's crown. The Sega Nomad was only released in the North American market, where Sega's attention was more focussed on the Saturn. Nintendo offered the Virtual Boy, which offered an underwhelming stereoscopic 3D experience that was uncomfortable to play for prolonged periods. Neither sold well.

And in 1996, a new category was born. Remember *Tamagotchi*? More than 80 million units later, you can still buy the digital pets created by Bandai.

By the sixth generation, the growth of consoles had permanently shrunk the arcade video gaming market. And two

Nintendo Virtual Boy (1995). *(Evan-Amos / Public domain)*

Creator of *Tamagotchi*, Aki Maita. *(Yoshikazu Tsuno/PA Archive/PA Images)*

new types of media began to mature: mobile and internet gaming. The first game released on a mobile phone was in 1994, a *Tetris* variant on the Danish Hagenuk MT-2000. But most people's first experience with mobile gaming was *Snake*, pre-installed on Nokia phones from 1998 onwards. Meanwhile, online gaming gained popularity with titles such as role-playing game *Ultima Online* in 1997 and space war game *Starcraft* in 1998, which both debuted on the Microsoft Windows operating system.

Sega sought to distance itself from its recent defeats, to the point where its new high-tech console, the Dreamcast, didn't even have the Sega logo on it. The Dreamcast was launched in Japan in 1998 and elsewhere the following year, initially to critical acclaim. Its in-built modem and web access hinted at the future of gaming, and its visually impressive games were open, complex and reactive to player curiosity. But the

Dreamcast was not supported by some of the major game makers such as Electronic Arts, and the hype surrounding the upcoming PlayStation 2 ultimately killed the system. After the Dreamcast was prematurely discontinued in 2001, Sega pulled out of the console manufacturing business for good, focussing instead on developing games.

Sony's PlayStation 2 (PS2), launched in 2000, was destined to become the best-selling video game console of all time (and the first to outsell the Game Boy), shipping over 159 million units by the time it was discontinued in 2013. It could play DVDs, and was fully backward-compatible with the original PlayStation. Over 3,800 game titles would be released for the PS2, collectively selling some one and a half billion copies.

Snake II (2000), on the popular Nokia 3310, was many people's first game played on a phone. (Nokia)

Sony PlayStation 2 (2000), and the Slimline version (2004) (Evan-Amos / Public domain)

God of War (2005), released on the Sony PlayStation 2, became a flagship PlayStation franchise. (SCE Santa Monica Studio / Sony Computer Entertainment)

The other two major sixth generation consoles were Nintendo's Gamecube, and the new guy in town – Microsoft's Xbox. They sold respectably (over 20 million each), despite eating the PS2's dust. Nintendo also had a very successful series of handhelds with the Game Boy Advance, Advance SP and Micro eventually selling over 81 million units; much more popular than Nokia's disappointing effort, the handheld N-Gage, which only sold 3 million.

Microsoft Xbox (2001) *(Evan-Amos / Public domain)*

Halo: Combat Evolved (2001) was a launch title for the Microsoft Xbox. *(Bungie / Microsoft Game Studios)*

Nokia N-Gage (2003). A numerical keypad made the layout awkward for gaming. *(J-P Kärnä / CC BY-SA)*

Call of Duty (2003) was released for the Nokia N-Gage as well as for personal computers. Later *Call of Duty* games found more success on mainstream console platforms. *(Infinity Ward / Activision)*

Handheld platforms peaked with the seventh generation. In 2004, Sony released the immensely popular PlayStation Portable, and Nintendo gambled with a lower-tech offering that had a novel interface: the Nintendo DS, with its hinged pair of screens. The gamble paid off – the Nintendo DS appealed particularly to young children, with games like the *Tamagotchi*-inspired *Nintendogs*, and to middle-aged adults with its brain training games. It ended up selling 154 million units, almost as many as the lofty PlayStation 2.

Nintendo's willingness to forego hardware superiority in favour of an innovative interface paid off in spades with the release of the Wii in 2006. Again, rather than trying to appeal to hardcore players (mostly males from 18–35), Nintendo extended the market to a broader demographic, with the movement-sensing controller allowing the console to function as fitness equipment and family entertainment centre as well as a gaming device. Atypically, Nintendo profited from the sale of each Wii (usually manufacturers accept a loss on the sale of the console hardware, aiming to make money on software).

Microsoft's follow-up to the Xbox was the Xbox 360, launched in 2005 – a year earlier than the Nintendo Wii.

PlayStation Portable (2004). *(Evan-Amos / Public domain)*

Nintendo DS (2004), the second-best selling console of all time after the PlayStation 2, and one of only six to have sold more than 100 million units (along with Game Boy/Game Boy Color, PlayStation 4, PlayStation and Wii). *(Omega21 / CC BY-SA 3.0)*

The Xbox 360 sold extremely well, but it wasn't long before the Wii, and Sony's PlayStation 3, caught up and overtook it. Both Sony and Microsoft released motion controllers in an attempt to emulate the Wii's success. Microsoft's Kinect dispensed with controllers altogether, allowing players to control the action with their bodies – and notably, after selling 8 million units in its first 60 days, earned a Guinness World Record for being the fastest selling consumer electronics device ever.

Trends in the seventh generation included the ability to download and play games that had been released for older generation consoles; and consoles competing with paid-for TV by offering the ability to download or stream TV content such as films and sports. Starting in 2010, a series of so-called 'microconsoles' hit the market, such as the Ouya, which raised $8.6 million in preorders on the crowdfunding platform Kickstarter. These microconsoles allowed users to download games and digital media using pre-existing digital distribution channels such as Android or Steam (read more about Steam in Level 3: People and Personalities).

As well as using a motion controller, the player could interact with a Nintendo Wii (2006) by standing on a 'balance board'. *(Sergey Galyonkin from Raleigh, USA / CC BY-SA 2.0)*

Above: Microsoft's Don Mattrick and filmmaker Steven Spielberg announced the Microsoft Kinect at the Electronic Entertainment Expo in 2009. The project was then codenamed 'Project Natal' after the Brazilian city. Microsoft's marketing budget for the Kinect was alleged to be half a billion dollars. *(Antonio Fucito from Terni, Italia / CC BY-SA 2.0)*

Left: Bioshock (2007) was released on the Xbox 360. *(2K Games)*

Below: Assassin's Creed (2007) was released on the PlayStation 3 and Xbox 360. *(Ubisoft)*

Ouya's 2012 Kickstarter campaign. *(Ouya, Inc./Kickstarter)*

In 2007, video game consoles represented about a quarter of global general-purpose computational power. Consoles had represented the largest share of the video game market since overtaking arcade video games in 1997, but in the 2010s that changed – not because the consoles market shrank, but because other platforms grew.

In 2011, with the proliferation of smartphones, revenue from mobile games overtook consoles, and continued to grow rapidly, doubling the size of the entire video game market by 2019. And in 2013, PC game revenue grew enough to push consoles into third place in terms of market share.

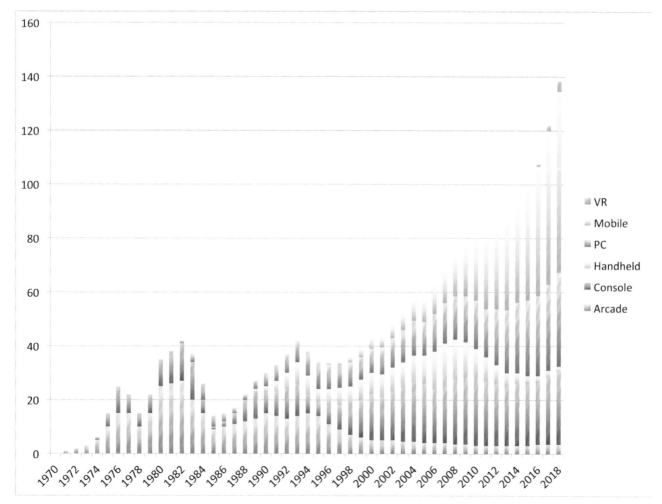

Video game industry global revenues by platform (billions of USD). This shows 1970–2018, after which analysts expect the graph to plateau. *(Charlie Fish / CC BY 4.0)*

With the spread of smartphones, suddenly the vast majority of the population had a gaming device in their pocket, offering a greater range of gaming experiences than any previous platform. In 2014, there were an estimated 1.8 billion video gamers worldwide – by 2018, 2.3 billion. A new type of game entered the market, free to play, comprising carefully calibrated microbursts of dopamine-driven satisfaction to get players hooked, then offering players the opportunity to buy tempting enhancements for a small charge. This model, typified by games like *Candy Crush Saga*, has proved dangerously profitable, encouraging game developers to make addictiveness their aim. In 2018 the World Health Organisation officially recognised 'gaming disorder' as a growing new kind of clinical addiction.

Candy Crush Saga (2012). *(Pexels from Pixabay)*

Meanwhile, the eighth generation of video game consoles started in 2012, with Nintendo's launch of the Wii U, which had mini screens built into its controllers. The Wii U did not sell well, but the technology was refined and incorporated into the much more successful Nintendo Switch, launched in 2017. The Switch was designed for both TV gaming and handheld gaming, further cannibalising a declining handheld market.

Both Nintendo consoles competed with Microsoft's Xbox One and Sony's PlayStation 4, each released in 2013. The pure handheld market was dominated by the first device

Nintendo Switch (2017). *(Evan-Amos / Public domain)*

Nintendo Wii U (2012). *(Benjamín Núñez González / CC BY-SA 4.0)*

Microsoft Xbox One (2013) with controller and Kinect sensor. *(Evan Amos / Public domain)*

to offer 3D gaming without needing stereoscopic glasses, Nintendo's 3DS. There was also a resurgence in virtual reality interfaces – the technology finally caught up to the idea – with units such as the Oculus Rift, which raised $2.4 million of crowdfunding on Kickstarter and was sold to Facebook two years later for $2 billion.

Digital distribution of video games had a big impact on the game development industry, allowing smaller games to reach a wider audience. This trend was bolstered by Kickstarter and other crowdfunding platforms, where small independent companies could raise money on the promise of producing bold, unconventional games; with

PlayStation 4 (2013). *(InspiredImages from Pixabay)*

An early development version of the Oculus Rift, which was released in 2016. *(Skydeas / CC BY-SA 4.0)*

enough success that the biggest publishers ultimately embraced smaller and self-published productions alongside their flagship blockbusters. An example of an 'indie' game that benefited from the backing of a major publisher is 2012's *Journey*, in which the player controls a silent robed figure crossing a desert towards a distant mountain.

Some of the latest technology that has inspired game developers includes voice and facial recognition, gesture control, and augmented reality. In the 2016 smartphone

Journey (2012). *(Thatgamecompany/Sony Computer Entertainment)*

Pokémon Go (2016). *(Tumisu from Pixabay)*

game *Pokémon Go*, the player swiped the screen to capture a creature superimposed onto their real-life surroundings through the phone camera. As of 2019 *Pokémon Go* had been downloaded over a billion times and there were about 150 million active players at any given time.

Recently, the video gaming market has come full circle, with a series of miniaturised retro gaming consoles being released, such as 2016's NES Classic Edition which comes with 30 built-in games from the licensed NES library, allowing adults to revisit their childhood favourites.

Technology has played a core part in the changing nature of video games, and fortunes have been won and lost in the race to create the next successful platform. Who knows what the future holds?

FIFA International Soccer (1993) on the Sega Mega Drive. A FIFA game has been released annually since. *(Extended Play Productions Creative Assembly / EA Sports)*

More than a quarter-century later, *FIFA 20* (2019). *(EA Sports)*

Level 3
PEOPLE AND PERSONALITIES

There are a handful of people without whom video games would not be what they are today. This chapter tells the stories of ten of the greatest video gaming visionaries.

RALPH BAER

Born: 8 March 1922, Rhineland-Palatinate, Germany
Died: 6 December 2014
Contribution: The Father of Video Games

In 2006, 83-year-old Ralph Baer was sitting in the East Room of the White House, alongside Roger L. Easton, the inventor of the Global Positioning System, and filmmaker George Lucas. They were among a select group of people waiting to be awarded the highest honour the US can confer upon citizens for contributions to technology, the National Medal of Technology. Baer, known as the 'Father of Video Games', was being recognised for having developed the world's first video game console, the Magnavox Odyssey, released in 1972.

Ralph Baer receiving the National Medal of Technology from George W. Bush in 2006. *(Eric Draper / Public domain)*

Baer was born in southwest Germany near the town of Pirmasens, which was famous for manufacturing shoes. His father ran a business tanning leather for the local shoe factories. But it was not a good time for a Jewish family living in Germany – in 1936, at age 14, Baer was expelled from his school, not for poor behaviour, but because of his ancestry. He was forced to attend an all-Jewish school instead.

Fearing further persecution from the Nazi party and its supporters, in 1938 the family emigrated via the Netherlands to the US, and settled in New York City. They were just in time – two months later was *Kristallnacht*, the Night of Broken Glass, during which 7,000 Jewish businesses across Germany were damaged and destroyed, and 30,000 Jewish men arrested.

Baer was drafted into the US Army in 1943, where he was trained in military intelligence and psychological warfare. During the war he learned to fix up radios, but his real interest was guns. He studied guns and collected them, and when the war was over he shipped 18 tons of foreign weapons back to the US, exhibiting them in museums in Maryland, Massachusetts and Kansas.

The rest of his career was spent as an engineer, mostly for defence companies, working on a diverse range of projects and research. As he became more senior, his inventiveness was increasingly given free reign, and by the end of his life he had over 150 patents in his name, many for gaming devices and interactive toys. One of the toys he created was 'Simon', with four big coloured buttons which lit up in a sequence that the player had to memorise and recreate. He was also instrumental in Coleco's entry into the video game market (read more about Coleco in Level 4: Companies and Capitalism).

But Baer's biggest contribution to video gaming was the world's first commercial home gaming console, the Magnavox Odyssey. Baer had several courtroom chances to assert his status as a pioneer – even though Magnavox had left the video gaming industry by 1984, their lawyers

Milton Bradley's 'Simon'. *(Shritwod / Public domain)*

Ralph Baer, in 2010, working on a reproduction of the 'Brown Box' prototype of the Magnavox Odyssey. (Rolenta / CC BY-SA 4.0)

relied on Baer's patents to force other video game makers to pay them a licence fee. Magnavox initiated legal proceedings against many of the major early video gaming manufacturers for infringing Baer's patents, and by 1998 had collected nearly $100 million. The key patented characteristic they most often relied on was that the games included a 'play-controlled hitting symbol'. With so many early video games directly influenced by the tennis game *Pong*, no wonder Magnavox's litigators were rubbing their hands together with glee!

NOLAN BUSHNELL

Born: 5 February 1943, Utah, US
Contribution: Founded Atari

There was another contender for the title of 'The Father of Video Games', one who almost single-handedly commercialised the fledgling video game industry. Nolan Bushnell was born in Utah in 1943, to a Mormon family. As a student of the University of Utah College of Engineering, he displayed an entrepreneurial streak – he made a schedule of events at the university, sold advertising space around the edges, and distributed it quarterly.

He claims to have lost his entire tuition fund in one ill-advised poker game, after which he got a night job to keep himself out of trouble. His first duty at Lagoon

Nolan Bushnell in 2013. (Tech Cocktail / CC BY-SA 2.0)

amusement park near Salt Lake City was running the 'Spill-the-Milk' stall where players could win a stuffed toy by knocking over milk bottles with a baseball. He was quickly promoted to manager of the games room.

During this time he saw one of the first video games, *Spacewar!*, and immediately thought he could make money by bringing it to the amusement park. But the game only ran on a computer the size of a double wardrobe, which cost about ten times the price of a house, so he filed the idea for later.

After graduating in 1968, Bushnell worked for Ampex Corporation in California, known for breakthroughs in audio and video recording technology. There he read about a new computer that cost less than $4,000, and he thought of *Spacewar!*. He figured if he could get the computer to run three or four terminals, he could make the game profitable at 25 cents per play. But the computer was too slow. With the help of colleague and friend Ted Dabney, he built dedicated circuits to run the game instead, which also ended up being much cheaper to make.

Bushnell and Dabney created a partnership called Syzygy in 1971, and persuaded arcade game manufacturer Nutting Associates to hire them to develop a simplified version of *Spacewar!*, which ended up being called *Computer Space*. Distributors didn't quite know what to make of this newfangled technology – they were concerned about reliability, and worried that hooligans might steal the TV right out of the cabinet. The game ended up selling about as many units as the average pinball machine, but failed to appeal to a mass audience.

Syzygy's next break was a deal to produce video games for Chicago-based pinball juggernaut Bally Midway, starting with a driving game. Bushnell and Dabney renamed their company Atari, after a term from the ancient board game Go, and poached an engineer they knew from Ampex called Al Alcorn.

In 1972, Bushnell heard about another company developing video games. He made sure to check out the competition, and attended a demonstration of the Magnavox Odyssey, which featured a primitive table tennis game called *Ping-Pong*. To test his newly employed engineer's skills, he asked Alcorn to make a version of that game, pretending to him that the idea for the game was Bushnell's own, and that it was to fulfil a (non-existent) contract with General Electric.

Alcorn threw himself into the job, implemented some clever quirks that improved the gameplay, and *Pong* was born. A *Pong* cabinet was installed in a local bar, Andy Capp's Tavern (now Rooster T. Feathers), to test its popularity with the punters. Bushnell tried to use the game to fulfil their obligation to Bally Midway, but they didn't want a two-player game. Meanwhile, the test unit proved so popular, Bushnell knew he had a hit on his hands, and decided to bet the farm. He used all of Atari's money to manufacture 11 more *Pong* cabinets, sold those and used the proceeds to make more, and so on until *Pong* had taken the world by storm.

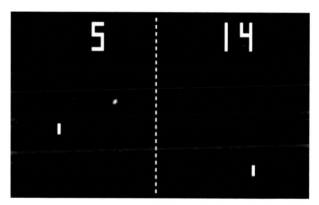

Pong (1972). *(Atari)*

Computer Space (1971). *(Flippers / Public domain)*

Atari founders Ted Dabney and Nolan Bushnell with Fred Marincic and Al Alcorn. *(Alcorn, Allan)*

While Bushnell and Dabney continued working together for another decade, there were many disagreements between them, not least when Dabney left Atari after Bushnell patented Dabney's video circuit idea without including him on the patent.

Bushnell became wealthy when entertainment giant Warner Communications bought Atari for $28 million, but was forced out of his own company in 1978 after a dispute with Warner over its future direction. Since then, Bushnell has created a series of businesses, notably the Chuck E. Cheese's Pizza Time Theatre chain of restaurants, and one of the earliest 'business incubators', Catalyst Technologies Venture Capital Group.

Chuck E. Cheese's Pizza. (Try to Love Again / Public domain)

Bushnell continues to be involved in the video game industry. For years he focussed on educational games, saying he wanted 'to leave a legacy of more than just fun'. In March 2019, he was appointed CEO and Chairman of Global Gaming Technologies Corp, which is capitalising on the growth of video games as a spectator sport with its Videre esports betting platform.

CLIVE SINCLAIR

Born: 30 July 1940, Surrey, England
Contribution: Kick-started the UK video games market

Nolan Bushnell wasn't the only early video game influencer who displayed a penchant for gambling. In 2003, Sir Clive Marles Sinclair won the first season of British television series Celebrity Poker Club – just one of his multifarious off-the-wall achievements, which also included chairing the high IQ society British Mensa from 1980 to 1997, and making a splash in British tabloids for marrying a lap-dancer 36 years his junior. Known as 'Uncle Clive', he has always cultivated his image as an eccentric, and became a household name in the UK after developing a series of hugely popular low-cost computers.

Born in 1940, Sinclair was one of the millions of children evacuated from urban areas at the start of the Second World War. He was precocious, with a talent for mathematics and electronics, and an inventive streak. At age 14, he designed

Sir Clive Sinclair in 1990, riding a prototype foldable X-Bike designed by Mark Sanders. (Mark Sanders / CC BY-SA 4.0)

a one-person submarine; by the age of 17, he was writing for *Practical Wireless* magazine and selling radio circuit kits by mail order.

He founded his first company, Sinclair Radionics, in 1961, focussed on creating innovative and inexpensive consumer-oriented products such as miniature transistor radios, pocket calculators, digital watches and micro TVs. By the end of the 1970s he sought to create an affordable home computer to compete with Apple.

The Sinclair ZX80 was launched in 1980, followed by the ZX81 a year later, using cheap components and minimalist design that kept the price well below £100 (£430 in 2020 pounds) – a vital ingredient for success given that the country was in a deep recession, and the target market was not businessmen but teenagers dreaming of making arcade games in their bedrooms.

But the real breakthrough machine was 1982's full-colour ZX Spectrum. Launched at half the price of its main rival, the BBC Micro, and a third of the price of the Commodore 64, it sold millions of units and inspired a generation of young British programmers. In 1982 alone, 226 British-made Spectrum games were released. In 1983, 1,188 were released – made by 458 different companies.

The kind of games that were made for the ZX Spectrum reflected the eccentricity of the computer's creator.

Sinclair ZX80 (1980). *(Daniel Ryde, Skövde / CC BY-SA 3.0)*

Sinclair ZX Spectrum (1982). *(Bill Bertram / CC BY-SA 2.5)*

Sinclair ZX81 (1981). Programs were loaded onto the Sinclair computers using an ordinary audio tape recorder. *(Mike Cattell / CC BY 2.0)*

People riding battery-assisted Sinclair C5s. *(Prioryman / CC BY-SA 4.0)*

Manic Miner (1983). *(Matthew Smith / Bug-Byte)*

The quirky British sense of humour was on full display in titles such as *Manic Miner*, made by 17-year-old Matthew Smith and featuring colourful surrealist delights such as deadly penguins, walking toilets and flying telephones, alongside cultural references such as ewoks and *Donkey Kong* parody levels. (More than 30 years after its release, *Manic Miner* was ranked by gaming website Polygon as one of the top 100 games of all time.)

Sinclair was knighted in 1983 for 'services to British industry'. His output as an inventor has continued to be prolific ever since, although often with less commercial success – he notoriously lost a small fortune with the Sinclair C5, a poorly marketed electrically assisted pedal cycle that was meant to revolutionise transport but instead bankrupted the company.

ROBERTA WILLIAMS

Born: 16 February 1953, California, US
Contribution: Popularised graphic adventure games

In 1979, Roberta Williams (née Heuer) was a 26-year-old housewife caring for two young children, when her husband Ken introduced her to the world's first text adventure game, *Colossal Cave Adventure* (released in 1977). They played it together, enjoyed it, and looked for others like it. Underwhelmed with what they found, Roberta Williams decided to create a game of her own. Inspired by Agatha Christie's murder mystery novel *And Then There Were None*, Williams started writing the story, and suggested including graphics. Her husband helped her with the programming, finding a way to compress the 70 scenes she had drawn onto a 5¼ inch floppy disk.

The resulting game, *Mystery House*, the first ever graphic adventure game, was a hit. The couple co-founded On-Line Systems and continued making graphic adventure games, renaming their company to Sierra On-Line after their success afforded them a new house in the foothills of the Sierra Nevada mountains.

By the end of 1982, the company had released 16 games, a handful of which had been designed by Williams (and she appeared in a hot tub on the cover art of one, *Softporn Adventure*). But her most lasting contribution to the canon of video gaming came in 1983, with *King's Quest*, a full colour adventure game inspired by fairy tales, in which the player controls an on-screen character to explore a large map, interacting with other characters and solving puzzles to progress the story.

Mystery House (1980). *(On-Line Systems)*

Ken and Roberta Williams in the early 1990s. *(John Storey/The LIFE Images Collection/Getty Images)*

Softporn Adventure (1981). Roberta Williams is on the right. *(On-Line Systems)*

The original *King's Quest* (1984) being played on an Apple IIe computer. *(Blake Patterson / CC BY 2.0)*

King's Quest spawned seven sequels and a host of spin-offs. Williams' design philosophy focussed on building beautiful worlds filled with characters that players could 'relate to and care about', and she innovated with icon-based interfaces (rather than parser-based) to make the adventures accessible to a wider audience.

The game she has claimed to be proudest of is 1995's *Phantasmagoria*, a full motion video 'interactive movie' in which the protagonist uncovers the murderous secrets of a remote New England mansion. Its horror theme was a major departure from her previous family-oriented games. *Phantasmagoria* was developed over the course of two years for $4.5 million (the original budget was $800,000); it had a script of over 500 pages and was published on seven CD-ROMs. Despite receiving mixed reviews, and courting controversy due to its depictions of gore and

Two *King's Quest* games for the Atari ST (a computer released in 1985). *(Tomer Gabel / CC BY-SA 2.0)*

Phantasmagoria (1995). *(Sierra On-Line)*

sexual violence, it was a commercial success, selling over a million copies.

Williams retired from making video games in 1999 to travel the world with her husband on their yacht. She hasn't been involved in the video game industry since, apart from reportedly helping out with 2013 Facebook game *Odd Manor*. Sierra On-Line was sold in 1996, but as part of other companies continued to publish games, such as 1998's popular first-person shooter *Half-Life*.

HENK ROGERS

Born: 24 December 1953, Amsterdam, Netherlands
Contribution: Brought *Tetris* to the world

Soviet-born Alexey Pajitnov famously created (with the help of two colleagues) one of the best selling games of all

Henk Rogers in 2010. *(Al Pavangkanan from Van Nuys, USA / CC BY 2.0)*

time, *Tetris*, but without Henk Rogers you might never have played it.

Rogers was born as Henk Brouwer, but when he was seven years old his mother married an American and they moved to New York, where he appended the surname Rogers. He learned English, and did a computer science elective at Stuyvesant High School before dropping out.

His American father was an avid board gamer, whose passion for the ancient game of Go drew him to Japan. But the family only made it halfway, to Hawaii, where Rogers took up computing at the University of Hawaii, paying his way by driving a taxi. But again, he never finished his studies – Rogers fell in love with a Japanese woman, and followed his father to Japan (that woman, Akemi, is still his wife four decades later).

In the early 1980s, as personal computers were becoming popular in Japan, Rogers made a roleplaying game inspired by Dungeons & Dragons. The Black Onyx ended up being the best selling game in Japan in 1984. He started a company called Bullet-Proof Software and travelled around the world looking for new games to publish. That's when he met Alexey Pajitnov.

Pajitnov was working on speech recognition at a government R&D academy in Moscow. In 1984, he decided to make a game to test the capabilities of the Electronika 60 computer. Inspired by pentomino puzzles he had played as a child, he made falling blocks out of text, since the computer had no graphics capability. Two weeks later, the prototype was done; it was satisfying to try and fit the pieces together, but the screen filled up too quickly, so he made completed lines disappear. The communist government promptly seized the rights and licensed the game for personal computers. Pajitnov didn't see a penny of profit.

The Black Onyx (1984) box cover art. *[Bullet-Proof Software]*

Soviet computers in 1985. *(Don S. Montgomery, US Navy (Ret.) / Public domain)*

Alexey Pajitnov in 2008. (Jordi Sabaté / CC BY 2.0)

Rogers saw the game's potential, and negotiated with the Soviet apparatchiks for the international publishing rights, befriending Pajitnov in the process. Rogers even made a couple of improvements to the game, such as adding a bonus for completing more than one line at once.

Previously, Rogers had taken advantage of his knowledge of Go to form a relationship with the all-powerful president of Nintendo, Hiroshi Yamauchi, himself a very strong Go player. At the time Yamauchi considered only the biggest arcade game publishers sophisticated enough to make games for Nintendo, but when Rogers offered to make a Go game for the NES, he accepted. The two men would later play Go together after business meetings.

Rogers leveraged that relationship to persuade Nintendo to bundle *Tetris* with every Game Boy sold in the US. Nintendo had been intending to bundle a Mario game – Rogers argued 'if you want little boys to buy your Game Boy, then include Mario. But if you want everyone to buy your Game Boy, then include *Tetris*.'

Rogers spent the next decade cornering all the rights to *Tetris*, which he vested in The Tetris Company, half of which was owned by the Soviet Ministry of Hardware and Software (ELORG), the other half split evenly between Rogers and Pajitnov. The Tetris Company made a fortune by getting *Tetris* onto mobile phones. In 2005, Rogers' mobile phone rights for *Tetris* were due to expire, and rather than renew them with ELORG, he bought ELORG (which had become a private business after the fall of the Soviet Union).

These days, Rogers lives in Hawaii. After a near-death experience in the form of a heart attack, he set up a charity called Blue Planet Foundation focussed on renewable energy, and is now dabbling in space exploration. Meanwhile, Pajitnov can be seen cruising the streets of Bellevue, near Seattle, in his Tesla Model S with the licence plate 'TETRIS'.

SID MEIER

Born: 24 February 1954, Ontario, Canada
Contribution: Legendary game designer

At the end of the 1980s a new type of game became popular. Up to that point, most games were products of the action-oriented coin-hungry philosophy of the arcades. But in 1989 Will Wright's city building simulation *SimCity*, and Peter Molyneux's god game *Populous* were both wide open sandbox games in which the player could decide their own objectives. Both Wright and Molyneux are legendary game designers in their own right – and they inspired another visionary designer to create his most enduring game: *Sid Meier's Civilization*.

Meier's mother was Dutch and his father Swiss. They intended to move to America, and stopped off in Canada

Tetris (1989) on the Game Boy. (William Warby from London, England / CC BY 2.0)

Sid Meier. (Official GDC / CC BY 2.0)

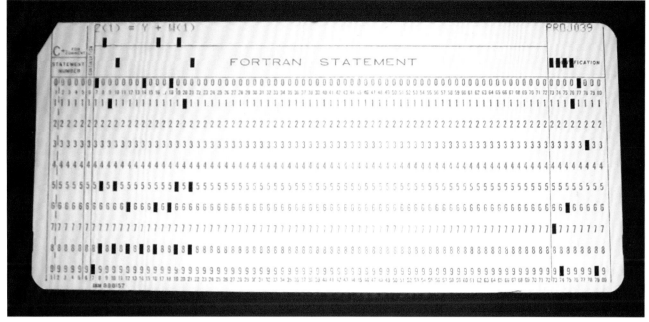

A Fortran program punched onto an 80-column card. (Arnold Reinhold / CC BY-SA 2.5)

to raise the $5,000 required to emigrate to the US. So Meier was born in Ontario in 1954 with dual Canadian and Swiss citizenship, before moving with the family to Detroit, Michigan.

As a child, Meier played with toy soldiers, and eventually graduated to Avalon Hill war games – complex strategic board games. He studied history and computing at the University of Michigan. At the time, mainframe computers processed data using decks of '80 column cards', essentially index cards with holes punched into them. He experimented with making simple games such as noughts and crosses (tic-tac-toe) on the format. But it was his purchase of an Atari 800 in 1981 that inspired him to make games to try and sell. He started making rip-offs of popular arcade games, programming them in 6502 Assembly machine language, copying them onto cassettes, and packaging them in Ziploc bags.

He was working for a company installing cash registers and the tote boards that display odds during horse races. During a trade show in Las Vegas, he met a colleague, 'Wild Bill' Stealey. The two of them started talking about video games, and found an arcade game called *Red Baron* in which you had to guide a biplane through abstract obstacles. Stealey, a former Air Force pilot, reckoned he could get a high score – but after watching Stealey play, Meier had a turn and got double his score. The pair decided then and there that they would make games together.

They founded Microprose in 1982, starting with mostly military simulations. Meier did the programming, art and sound; Stealey did the selling (he would call computer stores pretending to be different customers asking for the latest Microprose game until they were desperate to stock it). They had enough success that they were eventually able to hire artists, and even quit their day jobs.

Their big release in 1987 went in a new direction. It was also the first game that included Meier's name in the title, which would become a signature feature for Meier's games: *Sid Meier's Pirates!*

Despite Meier's interest in history and realism, he researched the game by reading children's books, to capture the essence of what made pirates interesting and fun. *Sid Meier's Pirates!* was an open game where you could explore the Caribbean by boat; decide to be a pirate or a pirate-hunter; ally with the British and then change your mind and side with the Dutch.

Sid Meier's Pirates! (1987). (MicroProse)

Sid Meier's Civilization (1991). *(MicroProse)*

XCOM: Enemy Unknown (2012). *(Firaxis Games / 2K Games)*

Meier exploited a technological trick wherein art could be folded into a font to be able to compress pictures more efficiently, allowing the game to be punctuated with pretty graphics.

Meier wanted to go bigger. His next major game was a business simulation called *Sid Meier's Railroad Tycoon*, which modelled the early development of railroad in the US and Europe. And after that, he wanted to go bigger still – so, he simulated the entirety of civilisation, from 4000BC through to the near-future, complete with exploration, warfare and diplomacy. *Sid Meier's Civilization* was a massive hit, spawning many sequels and spin-offs.

In 1993, Microprose was sold, and shortly afterwards Meier and two fellow designers founded a new company, Firaxis Games. While real-time strategy games were becoming the norm, Firaxis Games has continued to buck the trend by focussing on turn-based games such as *XCOM: Enemy Unknown*.

Today, Meier lives in Baltimore, Maryland, where he is still Creative Director of Firaxis Games, and plays organ for his local church.

SHIGERU MIYAMOTO

Born: 16 November 1952, Kyoto, Japan
Contribution: Most influential game designer

In 2008, Shigeru Miyamoto topped Time Magazine's list of the 100 most influential people of the year. Not just the most influential video game designer, although he certainly is, but the most influential person. His creations, including Mario, the most popular video game character of all time, have had a profound impact on the evolution of video gaming and become part of our collective cultural consciousness.

He was born in the rural town of Sonobe outside Kyoto in Japan, in 1952. As a child, he made his own toys out of wood and string, and explored the wooded mountains around his home – which would later inspire him to imbue his video game designs with a sense of exploration and discovery. He travelled far from home, to Kanagawa, to study industrial design, and tried to become a professional artist of Japanese manga comics. He still enjoyed making toys, as well as playing bluegrass music on banjo and guitar.

In 1977, his father, an English teacher, got him an interview with the president of Nintendo, then a small toy company. Miyamoto brought to the interview some colourful wooden coat hangers he had made, in the shape of animals, explaining that they were safer and more fun for

Shigeru Miyamoto in 2007. *(Sklathill / CC BY-SA 2.0)*

Radar Scope (1980). *(Nintendo Arcade)*

children than normal hangers. He got the job, starting as an apprentice in the planning department.

Over the next two years Nintendo branched out into arcade video games, and Miyamoto was helping with the art and design of the games – everything from drawing pixel images to designing the arcade cabinets themselves. Miyamoto's big opportunity came when Nintendo tried to break into the American market with a *Space Invaders*-inspired game called *Radar Scope*. Despite selling reasonably well in Japan, the game flopped in the US, leaving Nintendo with 2,000 unsold arcade units, and in 1981 Miyamoto was assigned to develop a game that would run on these units instead.

He wanted to make a game based on the love triangle between Popeye, Bluto and Olive Oyl, but Nintendo were not able to secure the rights to use the *Popeye* characters. So instead he used a carpenter called Mr Video, later renamed Jumpman, alongside a strawberry blonde damsel in distress in a pink dress called Lady, and a big gorilla inspired by King Kong. Jumpman had to climb a tower, dodging fireballs and rolling barrels, to rescue his girlfriend from the giant ape.

When the game was sent to the Nintendo of America warehouse, the characters were renamed to Mario (after the Italian warehouse landlord), Pauline (after the warehouse manager's wife), and Donkey Kong (Miyamoto chose 'Donkey' to evoke stupidity and stubbornness).

At the time, games like *Space Invaders* and *Pac-Man* did not concern themselves with story, so Miyamoto's simple narrative and the cheeky personality of his characters made the game unique. *Donkey Kong* was also the first platformer, with a character progressing through a level by jumping to avoid obstacles. *Donkey Kong* ended up being one of the best selling arcade video games of all time – and Miyamoto was just getting started.

Mario became a plumber, and gained a green-clad brother, in Miyamoto's 1983 game *Mario Bros.* which had

Donkey Kong (1981). Joshua Driggs (ZapWizard), bayo / CC BY-SA 2.0)

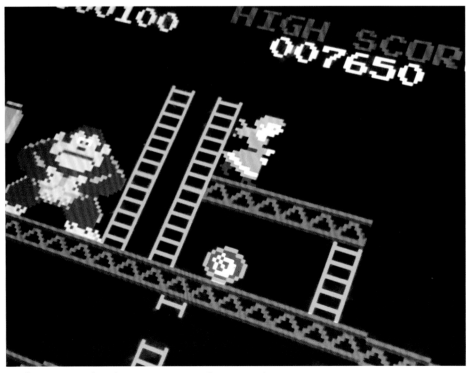

Donkey Kong and Pauline. (Microsiervos / CC BY 2.0)

the two characters clearing beasts out of the sewers. In 1985, the brothers went super – Miyamoto's *Super Mario Bros.* almost single-handedly dragged the US out of a stifling video game market crash, popularised side-scrolling platform games, and became one of the best selling games of all time.

His next big game was an expansive role-playing game in which the player explored to find a series of dungeons on a quest to rescue a princess: *The Legend of Zelda*. Mario and Zelda games have continued to be flagship titles for Nintendo ever since. Miyamoto has also been responsible for a host of other much-loved Nintendo titles

Characters created by Shigeru Miyamoto: Luigi, Yoshi and Mario. (Alexas_Fotos from Pixabay)

such as *Star Fox*, *Pokémon*, and pet simulation game *Nintendogs*, which was directly inspired by his Shetland Sheepdog Pikku, and his hobby of breeding dogs.

Miyamoto has a healthy suspicion of trends, preferring to do things his own way. For example, he strongly resisted putting in-game purchases into the mobile game *Super Mario Run* – players initially complained about the higher entry price point as a result, but that didn't stop over 300 million people from downloading a copy. He is humble about his

Nintendogs (2005) games for sale in Japan. *(Kevin Simpson / CC BY-SA 2.0)*

global impact, insisting that video games are highly collaborative projects. When out walking his dogs in Japan, he is more likely to be stopped by foreign tourists than Japanese locals, although he rarely signs autographs for fear of being inundated.

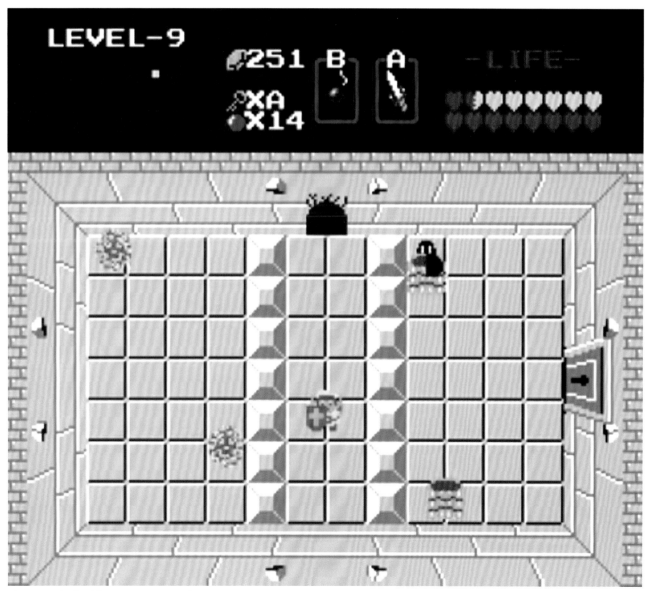

The Legend of Zelda (1986). *(Nintendo)*

Super Mario Run (2016). *(kyu3 / CC BY-SA 2.5)*

John Carmack in 2010. *(Official GDC / CC BY 2.0)*

He still enjoys playing bluegrass music on his banjo and guitar. He has been quoted as saying: 'Video games are bad for you? That's what they said about rock and roll.'

JOHN CARMACK

Born: 20 August 1970, Kansas, US
Contribution: Innovator of 3D graphics

John Carmack created new programming techniques that catalysed the rise in popularity of first-person shooters in the 1990s, in which the player sees the action through the eyes of the protagonist while blasting enemies away with a variety of weapons.

As a child, Carmack was, in his own words, an 'amoral little jerk'. From an early age his interests included computers, rockets, bombs, Dungeons & Dragons and arcades. In 1984, at age 14, he used a concoction of thermite and vaseline to burn through the windows of a school with the intention of stealing Apple II computers. Thanks to a clumsy accomplice, he was caught and sent for psychiatric evaluation.

The evaluation went very badly – he was described as having 'no empathy for other human beings' and as a 'brain on legs'. Despite this being a first offence, he was sentenced to a year in a juvenile home.

Later, he dropped out of the University of Missouri–Kansas City to start programming full time, but was unsatisfied with contract work, so he took a job for Softdisk, who made magazines published on a floppy disk, in Shreveport in Louisiana. He was finally happy, earning $27,000 a year, surrounded by experienced programmers he could learn from.

Three of his colleagues were fellow programmer John Romero, game designer Tom Hall, and artist Adrian Carmack (no relation). The four of them worked on a game, inspired by the console platform games popular at the time, using a technique John Carmack had developed allowing PCs to scroll smoothly. The game was published by Apogee Software as *Commander Keen in Invasion of the Vorticons*, as shareware – meaning that the first third of the game was given away for free via BBS bulletin boards, and players could send money if they wanted the rest.

The game sold well enough to quadruple Apogee's monthly sales, and the team's first royalty cheque topped $10,000 ($19,000 in 2020 dollars) – prompting them to quit Softdisk in 1991 and create their own company, id Software (eventually based in Dallas, Texas).

From the beginning, id did things differently. John Carmack worked his programming magic to come up with

Commander Keen in Invasion of the Vorticons (1990). *(Ideas from the Deep / Apogee Software)*

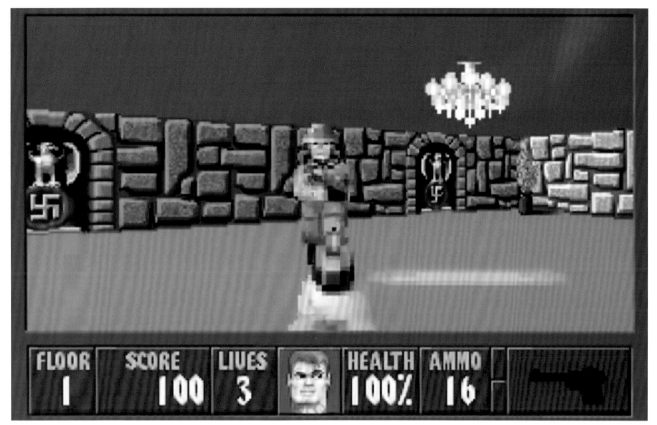

Wolfenstein 3D (1992). *(id Software / Apogee Software)*

a more powerful 3D graphics engine than had ever existed. One of their first games was 1992's *Wolfenstein 3D*, a first-person shooter in which the player navigated a maze-like underground bunker adorned with swastikas, blasting Nazis. It became the biggest selling shareware game up to that point, selling over 100,000 copies by the end of 1993.

Then, id did something that hadn't been done before – it offered to license Carmack's revolutionary 3D technology to other developers, so they could shortcut the process of developing an underlying software engine and focus on being creative instead. This 'middleware' licensing model ended up becoming standard industry practice

Buoyed by their success, id's next big game was even more ambitious, and even gorier. Carmack was determined to blow away the competition by rewriting his 3D graphics engine, and the team created a viciously dark, hellish game of destroying aliens with chainsaws and rocket launchers: *Doom*.

Doom (1993). *(id Software)*

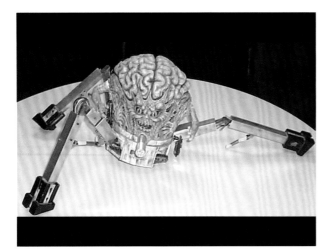

Model of the spider mastermind from *Doom* (1993), from the id Software offices. *(User Fredrik on en.wikipedia / CC BY-SA 3.0)*

Then, again, id did something new, by supporting fans to create their own content for the game. Up to that point developers had strongly resisted such activity as a breach of copyright, but id actively encouraged 'modding', a policy which successfully extended the shelf life of the game. They also took advantage of the growing popularity of the internet to allow players to compete simultaneously online in what id frontman John Romero dubbed 'death matches'.

After *Doom* came *Quake* in 1996, another enduring first-person shooter. John Carmack stayed with id, continuing

John Romero. *(Jason Scott / CC BY 2.0)*

to innovate new programming techniques and promote an open-source approach to software, until 2013, when he became Chief Technology Officer of Oculus VR, helping to develop

Quake (1996). *(id Software / GT Interactive)*

their virtual reality headset. He also revisited an old hobby from his youth, rocketry, twice winning $500,000 awards from the X Prize Foundation for building rocket systems.

KEN KUTARAGI

Born: 2 August 1950, Tokyo, Japan
Contribution: Father of the PlayStation

In the 1990s, Sony was cautiously considering entering the video game industry – but one man in particular was determined to make sure they didn't do things by halves. Ken Kutaragi's vision was to create a new console that would challenge Nintendo's stranglehold on the market, and he ended up creating the best selling console of all time.

As a child growing up in Tokyo, Kutaragi preferred to take apart his toys rather than play with them, and his parents encouraged his mechanical aptitude. He studied electronics at university, then went straight into a job in Sony's digital research labs in the 1970s.

He gained a reputation as a problem solver, and a hothead who would readily rat out his superiors, while working on cutting edge technology such as LCD screens and digital cameras. In the latter half of the 1980s he signed up to a contract with Nintendo to develop a sound chip for their upcoming Super Nintendo Entertainment System (SNES). At the time Sony had no interest in video games, believing them to be a passing fad, and his superiors would probably not have approved such a deal – so he went ahead in secret. When the project was discovered, only direct intervention from Sony's CEO Norio Ohga saved Kutaragi, and he was able to finish developing the sound chip, which ended up in every SNES.

Nintendo was so pleased with the sound chip that they came back to Kutaragi to develop a CD peripheral for the SNES. Sony was hesitant, but Kutaragi went ahead anyway and negotiated a deal very favourable to Sony – essentially, Sony would have full control of the SNES-CD format.

Ken Kutaragi receiving a Lifetime Achievement Award at the Game Developers Choice Awards 2014. *(Official GDC / CC BY 2.0)*

The SPC700 sound chip for the SNES was advanced for its time. *(Yaca2671 / CC BY-SA 3.0)*

The Nintendo PlayStation was planned as a stand-alone console as well as a CD-ROM peripheral for the Super Nintendo Entertainment System, but it was not to be. *(Paquitogio / Public domain)*

This rare Nintendo PlayStation prototype was sold at auction in March 2020 for £300,000. Pictured alongside a Sony PlayStation (1994). *(Mats Lindh / CC BY)*

Nintendo, notorious for always wanting to be in control, got nervous and very publicly dumped Sony.

Having backed this project against Sony's will, Kutaragi's career seemed doomed, but again Sony's CEO bailed him out. Kutaragi channelled Sony's rage at being slighted by Nintendo into an opportunity to create a whole new console, to directly compete with Nintendo. Sony Computer Entertainment was established, with Kutaragi in charge. (Technically, he was in charge of the engineering division, but he routinely got involved in business deals and marketing, sometimes to the chagrin of Steve Race, who was actually in charge.)

Kutaragi worked day and night, bullying his engineers to push the technology to its limits in a way that has seen him compared to abrasive visionary Steve Jobs. The PlayStation was the first (non-handheld) console to sell 100 million units.

After the PlayStation's phenomenal success, Kutaragi's ascendancy was assured; he became CEO of Sony Computer Entertainment America in 1997. He took a gamble with the PlayStation 2, investing $2.5 billion in research and development for the system. The gamble paid off, and the PlayStation 2 ended up becoming the best-selling console of all time.

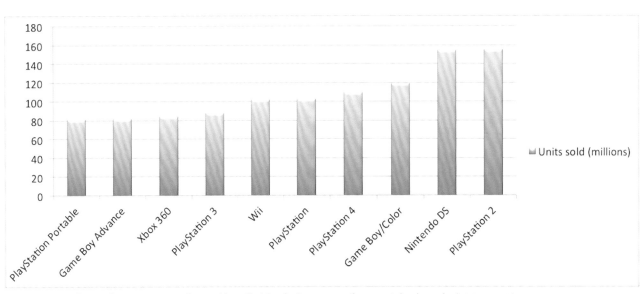

Top ten best selling video game consoles and handhelds of all time, as of 2019. *(Charlie Fish / CC BY 4.0)*

But Kutaragi had stepped on a lot of toes to get to where he was. When Norio Ohga stepped down as Sony's CEO in 1999, Kutaragi lost a vital ally; Ohga's successor Nobuyuki Idei was less tolerant of Kutaragi's brash manner. Idei promoted Kutaragi to head of Sony's entire consumer electronics division, but in doing so set him up to fail – the division was struggling, and Kutaragi had made enemies by openly criticising its failure to compete effectively with Apple's revolutionary iPod.

So, instead of becoming Sony's next CEO as he had hoped, and business insiders had expected, Kutaragi was busted back down to CEO of Sony's Computer Entertainment division and put in charge of the faltering launch of the PlayStation 3, already late and over budget. He made wild claims for the PS3, and attacked his competitors in the press, further cementing his reputation as 'Crazy Ken'.

Ultimately, Kutaragi was forced out of Sony; however, his legacy as a brilliant and visionary engineer lives on. Most recently, he has become involved in the development of self-driving cars.

GABE NEWELL

Born: 3 November 1962, Colorado, US
Contribution: Revolutionised the distribution of video games

Saturday 24 August 1996 was a big day for Gabe Newell. On the same day, he got married, and founded a company called Valve Software. Both were very successful ventures – he and Lisa Mennet Newell are still married, and at least one of their two children has taken after his father and is designing video games; and Valve made Newell a multi-billionaire and the richest person in the video game industry.

Newell was born in 1962, and got hooked on computers from an early age. He attended Harvard University, until

Gabe Newell in 2010. *(Official GDC / CC BY 2.0)*

Microsoft's head of sales persuaded him to drop out in 1983 and work for another Harvard drop-out, Bill Gates. Newell spent the next 13 years as the lead developer of the first three versions of Microsoft's Windows operating system, making him a 'Microsoft millionaire'. He also worked on porting id Software's *Doom* to Windows.

When id Software released *Quake*, and made the game's engine available for other developers to license, Newell and his colleague Mike Harrington left Microsoft to establish a game company called Valve. Their debut game *Half-Life*, built on a heavily modified version of the *Quake* engine, was published by Sierra in 1998, to critical acclaim for its immersive narrative, and Guinness World Record breaking commercial success. (In 2008 Guinness listed it as the 'Best-Selling First-Person Shooter of All Time'.)

Half-Life (1998). *(Valve / Sierra Studios)*

The valve in the lobby of Valve's headquarters in Bellevue, US. *(Claire Bateman)*

completed, someone stole it. On 2 October 2003, the source code for *Half-Life 2* was published on the internet – a catastrophic leak that threatened the future of the entire company.

Newell turned to the FBI, and then to the gaming community, to find the hacker that had stolen his crown jewel, with no success. But in February 2004, Valve received an email from someone taking responsibility for hacking into Valve's servers, and asking for a job. Newell played along, setting up a phone interview with the 21-year-old German man named Axel Gembe. Gembe was keen to show off his smarts, and readily spilled the beans about how he had pulled off the hack, although he claimed not to have been the person who published the source code online, and was remorseful about the trouble he had caused.

Newell offered him a second interview in Valve's offices in Seattle, where he intended to turn him in to the FBI. But German authorities got wind of the plan and on 7 May 2004 a team of police officers with automatic weapons turned up in Gembe's bedroom and arrested him.

Later that year, *Half-Life 2* was finally released, and Valve took the unprecedented step of forcing buyers to play the game through their new online distribution portal called Steam. The game was another mega-hit, and soon the Steam platform started selling games developed by other companies as well as Valve's own blockbusters.

By the end of 2005, Steam had 3 million members logging in every week to play games and get automatic upgrades. In 2019, that figure was 90 million active members, with Valve controlling around two-thirds of the market for downloaded PC games. And Valve's operating margin for downloaded games is double that of games sold at retail.

Over the next few years, Newell poured over $15 million of his own money into Valve, which included buying out the company's co-founder. Valve embraced the 'modding' community, allowing fans to modify the code of their games to create new situations, characters and scenery.

During its five years of development, Valve's sequel to *Half-Life* was one of the world's most hotly anticipated games. And then, shortly before *Half-Life 2* was

Valve has continued to develop its own games, with such notable successes as first-person puzzle

Half-Life 2 (2004). *(Valve)*

A sentry turret from *Portal 2*, in Valve's US headquarters. *(Claire Bateman)*

platformer *Portal* and multiplayer online battle arena game *Dota 2*. Recently, it has entered the virtual reality market. Newell himself, thanks to his portly frame and unabashed nerdy camaraderie, has become a much-loved figure in the hardcore gaming community. Notoriously, he has a collection of over 600 knives, and is a fan of the animated TV series *My Little Pony: Friendship is Magic*.

Level 4
COMPANIES AND CAPITALISM

Before *Pong* in 1972, there was no such thing as a video game industry, so most of the companies that built the foundations of the industry were not video game companies.

Nintendo was founded in Kyoto in 1889 by Fusajiro Yamauchi to produce and sell 'hanafuda' – playing cards carefully handmade from the bark of mulberry and mitsumata trees and printed with symbols inspired by nature. The word 'Nintendo' in Japanese means roughly 'leave luck to heaven'.

The cards became popular for gambling, allowing Yamauchi to expand his operations. In 1907, he successfully introduced Western-style playing cards to the Japanese market, distributing them via cigarette shops. Yamauchi's son-in-law Sekiryo Kaneda took over the company in 1929 and continued to grow the business, until 20 years later he suffered a stroke that forced him to retire. He summoned his 21-year-old grandson Hiroshi Yamauchi to his death bed in 1949, and asked him to abandon his law degree so he could take over the company.

Nintendo name plate on the original headquarters in Kyoto, Japan. *(Arcimboldo / CC BY 2.5)*

Nintendo poster from the late Meiji era. *(Public domain)*

Nintendo Western-style playing cards featuring Disney characters. *(Dinky Dana)*

Hiroshi Yamauchi, over the course of a 53-year tenure, was destined to transform Nintendo into a multi-billion dollar video game publisher and global conglomerate. In the beginning, he proved an ambitious and ruthless president, firing any employees who resisted his authority. His first big success was in 1959 when he struck a deal to use images of Disney characters on playing cards.

Determined to expand the scope of the company, Yamauchi experimented with a variety of business ideas including instant rice, a taxi service, and a love hotel with rooms rented by the hour (which he reportedly frequented). The failure of these ventures threatened financial disaster for the company, until Yamauchi realised that he could capitalise on Nintendo's existing distribution network into department stores for its playing cards, by manufacturing toys.

Yamauchi spotted a factory engineer, Gunpei Yokoi, playing with an extendable claw he had created to amuse himself – and he ordered Yokoi to develop it into a proper product. The 'Ultra Hand' was a huge success, and Yokoi was put in charge of new product development, creating a variety of innovative toys. Yokoi's first electronic product was a Love Tester machine that also sold well outside of Japan. In 1975, Nintendo entered the flourishing video game business and never looked back. Yokoi continued designing new products for Nintendo until 1996, his most famous contribution being the Game Boy.

Among the toys designed by Gunpei Yokoi for Nintendo was this 'Ultra Machine' from 1967, which pitched balls for batting. *(Vinelodge / CC BY-SA 3.0)*

Hiroshi Yamauchi in 1997. He died in 2013 at the age of 85. *(The Asahi Shimbun/Getty Images)*

Coleco was another foundational video game company that started life as something else. Founded by Maurice Greenberg in 1932 as the Connecticut Leather Company, it originally supplied leather to shoe companies, and branched out into selling rubber footwear in 1938. During the Second World War, Coleco benefited from the growing demand for the basic supplies the company produced, so that by the 1950s Greenberg was looking to expand. The success of a leather moccasin kit at the 1954 New York Toy Fair inspired Greenberg to focus on manufacturing toys.

The company invested in emerging technology for the vacuum forming of plastic, and produced a wide variety of plastic toys and paddling pools. By 1971, Coleco was listed on the New York Stock Exchange. Greenberg's son brought the company into the burgeoning video game industry in 1976 with the Telstar series of home consoles. Coleco had great success with handheld electronic games featuring licensed video arcade titles such as *Donkey Kong* and *Ms. Pac-Man*, and with 1982's ColecoVision home console.

Despite an unsuccessful venture into home computing with the Coleco Adam, the company initially weathered the 1983 video game crash comfortably thanks to the phenomenal success of its newest product, Cabbage Patch Kids dolls – but once the public grew tired of Cabbage Patch Kids, Coleco fell into a permanent decline and filed for bankruptcy in 1988. However, the Coleco brand remains alive and associated with video games to this day.

Sega's origin story begins in 1940, when three American businessmen in Hawaii founded Standard Games to provide slot machines to military bases. After the Second World War, the founders established a new company, called Service Games to highlight the military connection. When the US government outlawed slot machines in 1951, Service Games started providing slot machines and jukeboxes to US military bases in Japan, eventually expanding to South Korea, the Philippines and South Vietnam – and their machines began featuring the abbreviated 'Sega' brand.

Coleco made several consoles, including this Frankenstein's monster from 1977, the Coleco Telstar Arcade. *(Evan-Amos / Public domain)*

The Diamond 3 Star, a model of coin-operated slot machine produced by Sega in the 1950s. *(Rodw / Public domain)*

A promotional flyer for *Periscope* (1965). *(Sega Enterprises, Ltd.)*

Service Games was investigated for bribery and tax evasion, and eventually banned from US air bases in Japan and Philippines, leading to the dissolution of the company in 1960.

Meanwhile, a United States Air Force officer stationed in Japan called David Rosen founded a company in Tokyo in 1954 selling two-minute photo booths, and later importing coin-operated games. The successor company to Service Games bought Rosen's company in 1965, and Rosen was put in charge of the newly named Sega Enterprises.

Under Rosen's leadership, Sega transitioned into manufacturing. Rosen sketched out an idea for an electromechanical submarine game, which was released as *Periscope* in 1965. It was a huge hit, exported to Europe and the US. In the US, it became the first widespread game to charge 25 cents per play, which became the standard cost to play an arcade game for decades to come.

Sega released its first arcade video game in 1973 (a *Pong* clone, of course), and became one of the top arcade game manufacturers active in the United States. Sega's 1979 game *Head On* had cars racing around a maze to 'eat the dots', directly inspiring *Pac-Man*. In 1981, Sega licensed *Frogger* from Japanese developer Konami, which became Sega's most successful title to that point. And in 1982, Sega entered the home console market – with variable success over the next 20 years. At its peak, Sega eclipsed Nintendo as the most successful video game company, but by the end of 1990s the poor commercial performance of a series of flagship products cast Sega into serious financial difficulty.

Right: Sega's *Head On* (1979), precursor to *Pac-Man*. *(Sega Enterprises, Ltd./Gremlin)*

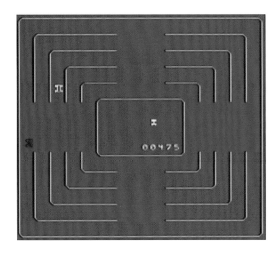

Below: *Frogger* (1981). *(Konami/Sega Enterprises, Ltd.)*

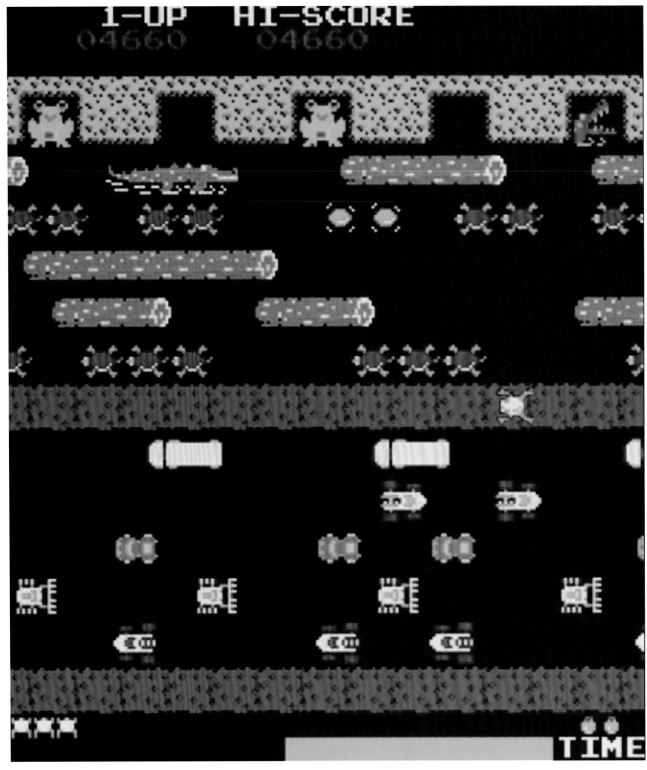

Since 1984, the majority shareholder in Sega was Japanese businessman Isao Okawa. His death in 2001 saved Sega. In his will, he forgave $40 million of debts Sega owed him, and gifted Sega $695 million of stock. Since then, Sega has stopped manufacturing consoles and instead focussed on developing and publishing games for third party platforms.

Apart from Nintendo, Coleco and Sega, there are only a handful of other companies still active in the video games industry that started before video games existed. Taito started out in 1953 importing vodka, vending machines and jukeboxes into Japan – and went on to make *Space Invaders*, *Bubble Bobble*, and many other much-loved games (Taito continues to make games, now as a wholly owned subsidiary of Square Enix). Bandai, which started as a spin-off from a textile wholesaler, was founded in 1950 to make toys – in the late 1990s Bandai nearly merged with Sega; but instead it merged with Namco in 2005 to become one of the world's biggest video game companies. Namco (originally short for Nakamura Manufacturing Company) started in 1955 as a producer of coin-operated department store children's rides. Konami was founded in 1969 as a jukebox rental and repair business in Osaka, Japan, by Kagemasa Kōzuki, who remains the company's chairman – Konami is responsible for many popular games including the *Castlevania*, *Metal Gear* and *Silent Hill* franchises.

Two of the first companies created specifically to produce video games were Atari and Electronic Arts, both in the US. The story of how they each started out provides an insight into how money has shaped the video game industry.

Nolan Bushnell and Ted Dabney's Atari was a scrappy, opportunistic company, created on the back of the game that launched an industry: *Pong*. Having failed to persuade Bally, Midway and Nutting Associates to manufacture *Pong*, Atari decided to go solo, using all of its meagre capital assets to make the first few machines in 1972. The bet paid off, and demand for *Pong* within the arcade business grew quickly.

Atari's manufacturing line was initially chaotic, with workers sourced from the local unemployment office, and many of the machines failing quality checks. The company threw parties with free beer (and incidental marijuana) if employees met their quotas, which encouraged an atmosphere of hard work and hard fun, and gradually the process was streamlined.

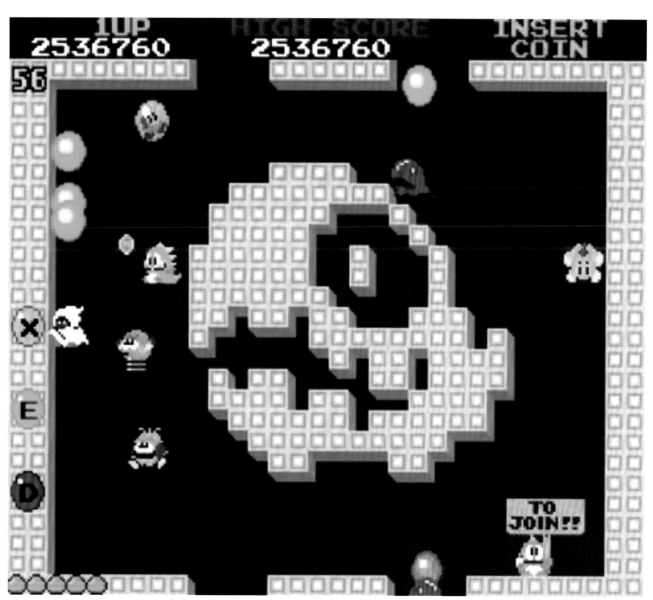

Bubble Bobble (1986). *(Taito)*

Castlevania (1986). *(Konami)*

The devotion of its staff helped Atari stay ahead of the curve even as much bigger manufacturers muscled in on their territory with countless *Pong* clones, some of which were as successful as the original. Atari experimented with new kinds of games like *Qwak!*, in which players shot ducks with a light gun, and *Gotcha*, in which players chased each other around a maze using pink domed controllers designed to look like breasts.

Characters from various Konami games. *(Sergey Galyonkin / CC BY-SA 2.0)*

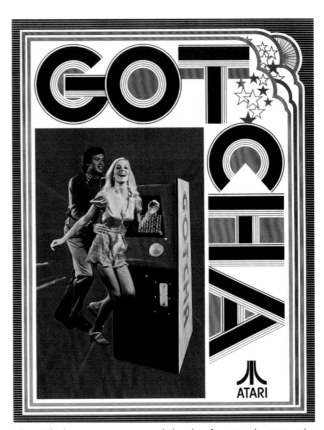

The pink domes were removed shortly after *Gotcha* (1973) hit the market. A limited run multicolour version of *Gotcha* was one of the first ever colour arcade video games. *(Atari)*

Gran Trak 10 (1974). *(Matt Griffin)*

In 1974 Atari created the first driving video game, *Gran Trak 10*, which became their best seller since *Pong*, but due to an accounting error the company underpriced the machine and lost $100 on every sale. This coincided with being sued by Magnavox for patent infringement (Atari eventually had to pay a punishing licence fee), and a failed attempt to enter the Japanese market by buying a Japanese factory with cash and no permits. Atari Japan was eventually sold to Nakamura Manufacturing, which became Namco.

Having lost $500,000, Atari was pulled back from the brink of financial disaster thanks to the success of a competitor's game: Kee Games' *Tank*, released in 1974, which outsold even *Pong*. Except Kee Games didn't really exist. Bushnell had created Kee Games to repackage Atari's games, so the same game could be sold to competing distributors who each demanded exclusive distribution deals. To maintain the illusion, Kee Games had its own offices, and Bushnell floated the rumour that Atari was suing Kee Games for theft of trade secrets. Using this cunning ploy, Atari was able to sell to twice as many distributors as its competitors.

After *Tank*, Kee Games was formally merged with Atari, which provided Atari the funds it needed to manufacture

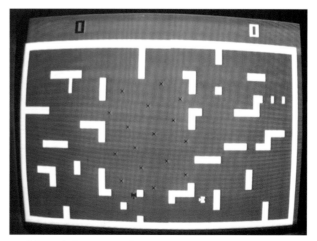

Tank (1974). *(Kee Games)*

Home Pong (1975). *(Evan-Amos / CC BY-SA 3.0)*

a home console version of *Pong*. But retailers were not interested in *Home Pong* – at $99.95 it was considered too expensive (that's about $480 in 2020 dollars). In 1975, as a 'last resort', Bushnell contacted the department store chain Sears Roebuck, at the time the largest retailer in the US, and suggested *Home Pong* for the sporting goods department. Bushnell estimated they could manufacture 75,000 units by Christmas – Sears asked for 150,000. Bushnell said 'Yes' without any clue of how he would meet the order (in the end, venture capitalist Don Valentine provided the funding).

In 1976 Bushnell sold Atari to Warner Communications for $28 million. Atari grew until it formed the largest part of Warner Communications. In 1982, Warner Communications spent $75 million promoting its products, more than Coca-Cola and McDonald's. But under CEO Ray Kassar's leadership Atari became less innovative and more focussed on licensed games.

That year, Atari was in talks with filmmaker Steven Spielberg for the rights to make a game based on *E.T. the Extra-Terrestrial* (at the time, the highest grossing film ever). The game was to be released in time for Christmas 1982, but talks only concluded in July. Atari paid around $22 million for the licence, and with a 1 September manufacturing deadline, there were only five weeks left to develop the game from scratch. The task fell to Howard Warshaw, who made a heroic effort to get the game made on time – but, inevitably, the final product didn't impress the critics and Atari was left with a mountain of unsold stock. That December, CEO Ray Kassar dumped 5,000 shares of Atari stock minutes before a press release went out announcing Atari's fiscal performance was 'substantially' below expectations. A few months later, Atari quietly buried 700,000 unsold game cartridges under concrete in a New Mexican landfill – a measure so extreme that it was considered an urban legend until the cartridges were exhumed in 2013.

In 1984, Atari was sold to Jack Tramiel (founder of home computer manufacturer Commodore International) for $50 and a stack of promissory notes. Atari persists as a video games brand to this day, and has come full circle with the release of a new Atari VCS console, with 'tons of classic Atari retro games pre-loaded, and current titles from a range of studios'.

In contrast, the origin story of Electronic Arts is considerably more organised and purposeful. California-born Trip

Exhumed Atari game cartridges in Alamogordo, New Mexico, US. *(taylorhatmaker / CC BY 2.0)*

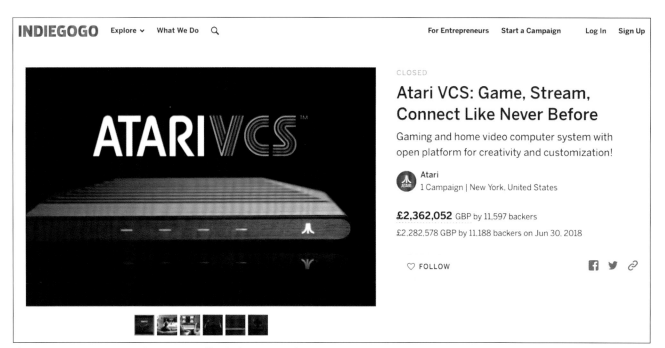

The 2020 Atari VCS console was crowdfunded. *(Atari, Indiegogo)*

Hawkins was a fan of sports strategy tabletop games as a child in the early 1960s, including *All-Star Baseball* and *Strat-o-Matic*. He envisioned a future where computers could make such games more accessible: 'as soon as I heard about computers, I could kind of see it with my own eyes… put all the administrative operating stuff in a box… [and] put pretty pictures on the screen like a television.' Thus began a two-decade quest to realise his dream.

Hawkins believed that 1982 would be the year that computer technology would catch up with his vision, so in 1975 he wrote a seven-year plan to capitalise on the video gaming revolution. He stuck doggedly to his plan. He tailored his undergraduate major at Harvard University in Strategy and Applied Game Theory so he could learn how to make video games. He took an MBA to learn the business skills he needed, and conducted market research on computer and console users. In 1978 he got a job at the Apple Computer Company, and made enough money after Apple went public in 1980 to fund his dream.

He quit Apple on 1 January 1982. With the help of venture capitalist Don Valentine, and a personal investment of over $200,000, he founded Electronic Arts (briefly called Amazin' Software). Apple's Steve Wozniak was a founder board member. Hawkins envisioned a Hollywood-style studio system for developing video games, with software designers as the stars. He was determined to sell directly to retailers, gaining an edge on competitors with higher margins and better market awareness.

In 1984, Hawkins approached American football coach John Madden to endorse a proposed video game. Madden turned out to be demanding, insisting that the game was an authentic simulation – to the extent that it took four years to finally release *John Madden Football*. Electronic Arts has since released a new version of *Madden NFL* (as it is now called) every year since 1990.

Hawkins' ambition for Electronic Arts games meant that he was keen to embrace more advanced 16-bit technology, so when Sega released the Mega Drive in Japan in 1988 he wanted to develop games for the platform. But Sega did not consider Electronic Arts an important enough company, and refused to send them a developer kit. Hawkins, determined, imported a Mega Drive from Japan and painstakingly reverse engineered it. He was prepared to develop games for the platform without an official licence, even if it meant being sued by Sega. In anticipation of being sued, Electronic Arts raised a contingency fund by taking the company public in 1989. However, Sega was so nervous that other companies would follow suit and undermine

Trip Hawkins in 2015. *(Christopher Michel / CC BY 2.0)*

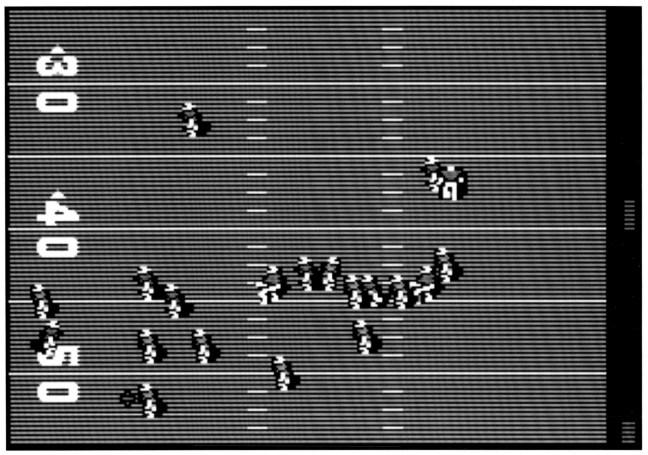

John Madden Football (1988). *(Robin Antonick / Electronic Arts)*

the licensing system, that they struck a deal – Electronic Arts would pay Sega $2 for each of the first million game cartridges sold, to become an official licensee.

Hawkins left Electronic Arts in 1991 to create a new games console with the innovative idea of licensing the console design to third parties. Meanwhile, thanks to the strong foundations Hawkins built, Electronic Arts grew into one of the largest video game companies in the world.

Any attempt to rank the world's biggest video game companies is necessarily an approximation, but as of 2019 these are among the largest.

Tencent (China)
Sony (Japan)
Microsoft (US)
Apple (US)
Activision Blizzard (US)
Google (US)
Netease (China)
Electronic Arts (US)
Nintendo (Japan)
Bandai Namco (Japan)

As of 2019, Rockstar North was the most successful UK games developer based on boxed game sales, out of over 2,200 UK video games companies. *(Rockstar North)*

Level 5
GENDER AND REPRESENTATION

In Shigeru Miyamoto's 1981 game *Donkey Kong*, the player must rescue a damsel in distress – a slender, pink-clad sprite called Pauline who has been kidnapped by the titular giant ape. Four years later, Miyamoto's *Super Mario Bros.* would become one of the best-selling video games of all time – it featured the quintessential damsel in distress, Princess Peach (also known as Princess Toadstool), whose in-game function was effectively to be the trophy.

In 1986, Miyamoto's *The Legend of Zelda* launched another immensely successful franchise in which the main female character, Zelda, despite having more agency than her counterpart Princess Peach, is inevitably kidnapped, cursed, possessed or otherwise disempowered and must rely on the male hero, Link, to rescue her. Zelda herself has never been a playable character in any of the main Zelda games, of which there are now over a dozen.

Side art on the arcade cabinet for *Donkey Kong* (1981), showing Pauline getting kidnapped. *(The International Arcade Museum)*

The ending of *Super Mario Bros.* (1985). *(Nintendo)*

Zelda is kidnapped again in *The Legend of Zelda: Skyward Sword* (2011). *(Nintendo)*

These games helped set the tone for gender representation in video games for decades: video games almost exclusively featured a male playable protagonist, with women generally presented as helpless or disempowered, as a motivation or reward for the player, or as background decoration, rather than as developed characters in their own right. Nintendo didn't invent the trope, but the popularity of their save-the-princess formula was hugely influential. That doesn't make those games (and others discussed in this chapter) bad – this chapter discusses some problematic aspects of certain games without casting judgment on those games as a whole.

Throughout the 1980s and 1990s, the primary consumers of video games were assumed to be young (straight, white) males. Sexualised representations of women, and variations on the damsel in distress plot, were often used as a lazy way to tap into male power fantasies to sell more games to boys and men. This attitude is self-perpetuating, with video games appealing disproportionately to males; and, as old games with regressive representations of female characters get rereleased and updated for new audiences, the pattern of presenting women as weak, ineffective or incapable is reinforced. As graphics have improved and games have become more sophisticated, the problem of sexism in video games has escalated, with women routinely hypersexualised (men rarely get the same treatment), and too-often existing only as disposable objects or symbolic pawns in stories about men.

One early video game specifically designed to appeal to female audiences was 1980s smash hit *Pac-Man*. Designer Toru Iwatani explained his thought process: 'When you think about things women like, you think about fashion, or fortune-telling, or food or dating boyfriends. So I decided to theme the game around "eating" – after eating dinner women like to have dessert.' Not the most enlightened logic, but thankfully no sexism was reflected in the content of the final game. The phenomenal success of *Pac-Man* in North America inspired a popular sequel, *Ms. Pac-Man*, which remains the most successful American-made arcade video game. *Ms. Pac-Man* featured one of the first female playable video game characters, and kicked off a trend of creating female versions of established male video game characters, generally by sticking a pink bow on them.

The pervasive habit of signifying female video game characters using a bow, pink clothing and exaggerated make-up is problematic in itself, not only because it perpetuates narrow and anachronistic ideas of femininity, but also because it emphasises that the default gender of video game characters is male. See, for example, the pink-bowed female characters in *Bubble Bobble* and *Adventures of Lolo*, and Wendy O. Koopa from the Mario franchise. Male characters are almost never specifically marked out as such, and correspondingly are portrayed in a much more diverse range of shapes and styles.

This gender imbalance is often aggravated by a game's marketing. Way back in 1971, the first commercially sold video game, *Computer Space*, despite being entirely gender-neutral in its content, was marketed with posters showing a nightie-clad woman leaning against the game cabinet. This tendency to use scantily-clad women in video game

Arcade cabinet art from *Ms. Pac-Man* (1982). *(Peter Burka / CC BY-SA 2.0)*

Lolo and Lala on the logo of *Adventures of Lolo 3* (1991). *(HAL Laboratory)*

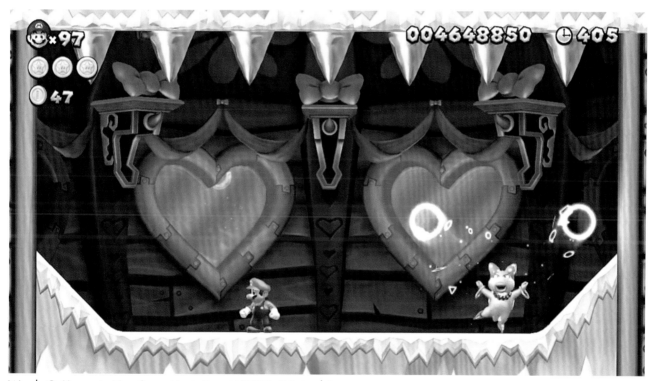

Wendy O. Koopa in *New Super Mario Bros. U* (2012). *(Nintendo)*

advertising has persisted. Rare's 2000 game *Perfect Dark* was notable for starring a female protagonist as its main playable character, secret agent Joanna Dark – but the television ad for the game focussed not on her skills but instead showed her getting dressed, and implied that her most important decision was what to wear for work. These marketing tactics seem to be less about selling the game, and more about selling a lifestyle where women exist as an object of heterosexual male desire. There has been a long-held perception that games with female protagonists do not sell as well – a study of games on the market in 2012 showed this to be true, but cited the fact that female-led games received less than 40 per cent of the marketing budget of their male-led counterparts, creating a vicious circle.

Video game developers have often resorted to tokenism in their representation of female characters. The much-loved *Mega Man* franchise featured 78 'Robot Masters' in its first 10 games, all male except for one: Splash Woman in the 2008 title *Mega Man 9*. The 2012 game *Scribblenauts Unlimited*, a game in which the player draws objects on screen to solve puzzles, features a family of 42 children (including the playable character), of which only one is

Computer Space (1971) poster. (Syzygy Engineering / Nutting Associates)

a girl. In the monolithic Mario franchise, Toadette seems to be the only female of her entire species. Game developer Ubisoft's questionable excuses for the absence of women in their 2014 games *Assassin's Creed: Unity* and *Far Cry 4* sparked an angry Twitter storm centred around the hashtag #WomenAreTooHardToAnimate. American essayist Katha Pollitt summed up the problem in a 1991 *New York Times* article: 'The message is clear. Boys are the norm, girls are the variation; boys are central, girls peripheral; boys are individuals, girls types. Boys define the group, its story, and its code of values. Girls only exist in relation to boys.'

Recent studies have shown that girls and women now make up half of all active videogamers, but still only about 4 per cent of games are being designed with women in a leading role.

Scribblenauts Unlimited (2012). (5th Cell / Warner Bros. Interactive Entertainment)

Still (with caption) from the British TV ad for *Perfect Dark* (2000). (Rare)

Super Mario Party (2018), with Toad and Toadette in the middle. *(Nintendo)*

The problem of sexism in video games mainly applies to first-person shooters, action games and sports games – in 2012, these three genres were responsible for 59 per cent of video game sales in North America. However, there are and always have been a significant proportion of games that avoid the issue of gender altogether, and often appeal more to female gamers as a result – such as *Tetris, Minesweeper, SimCity, Myst, Words With Friends* or *Candy Crush Saga*.

There are a handful of games in those three problematic genres (first-person shooters, action games and sports games) that have bucked the trend. The classic example is Nintendo's 1986 game *Metroid*, which became a long-lasting franchise. The protagonist of *Metroid* is space-faring bounty hunter Samus Aran, clad in stylish full-body armour. At the end of the game, it is revealed that Aran is not a man, as gamers had assumed, but a woman. A leap forward for female empowerment in video games; although it's unfortunate that the revelation of Aran's gender, depending on how quickly the player completed the game, was achieved by showing her in her underwear. This motif of rewarding players with women's bodies has persisted, with game series such as *Metal Gear, Resident Evil, Dead or*

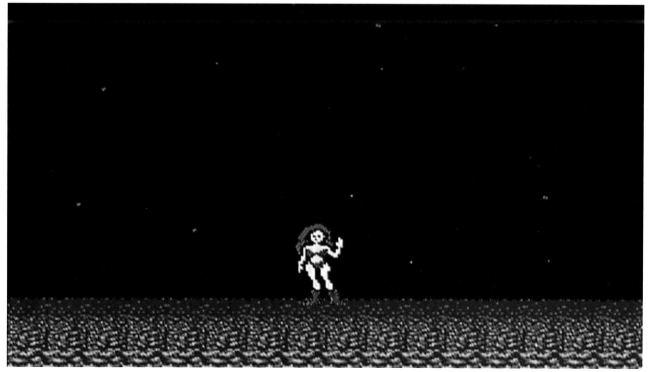

The faster the player completed *Metroid* (1986), the fewer clothes Samus Aran wore at the end. *(Nintendo)*

Metal Gear Solid V: The Phantom Pain (2015). *(Kojima Productions / Konami)*

Soulcaliber IV (2008). *(Project Soul / Namco)*

Alive and *Soulcaliber* featuring unlockable or downloadable content that includes skimpy outfits for female characters.

Some games have actively subverted the damsel in distress trope, such as the 1990 graphic adventure *Secret of Monkey Island*, in which the hero Guybrush Threepwood attempts to rescue the abducted governor Elaine Marley, but ends up spoiling her own escape plan. Or the 2008 indie release *Braid*, in which it becomes increasingly clear as the game progresses that the abducted princess has not actually been abducted at all, but is actually trying to escape from the playable main character.

In the latter half of the 1990s, the 'girl power' movement – embodied by the TV series *Buffy the Vampire Slayer* and British pop band Spice Girls – influenced video games, most notably with the 1996 release of *Tomb Raider*, which starred a kick-ass (if somewhat sexualised) female adventurer, and spawned a franchise. Other games released around that time with playable female protagonists include *Balloon Kid*; *Rhapsody: A Musical Adventure*; *Drakan: Order of the Flame*; *Kya: Dark Lineage*; *Primal*; and *Beyond Good & Evil*.

In recent years there has been a welcome growth of positive representations of women in video games,

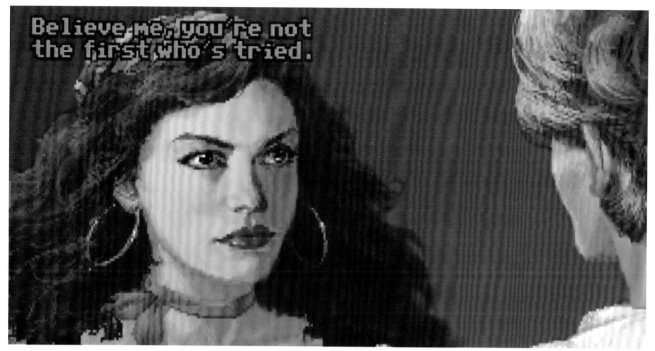
The Secret of Monkey Island (1990). *(Lucasfilm Games)*

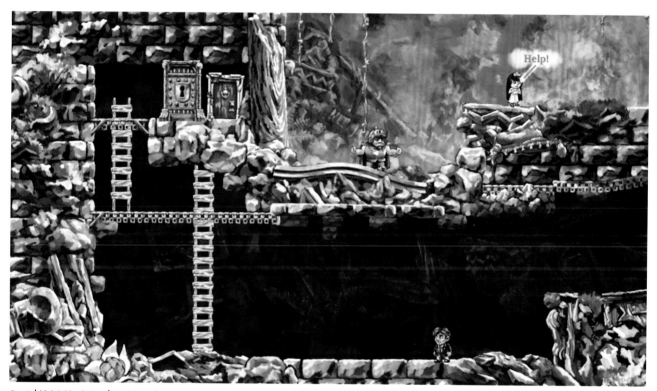
Braid (2008). *(Number None)*

Lara Croft from *Tomb Raider* (1996). *(Core Design / Eidos Interactive)*

Super Mario Odyssey (2017). (Nintendo)

including such blockbusters as 2008's *Mirror's Edge*, 2011's *Superbrothers: Sword & Sorcery EP*, 2012's *Gravity Rush* and 2013's *The Last of Us* – the latter of which is also notable for its positive representation of homosexual characters, another rarity in the history of video gaming. Coming full circle, the major 2017 Mario title *Super Mario Odyssey* features a more cosmopolitan and empowered Princess Peach, finally fed up of being fought over by Mario and Bowser.

In *Portal* (2007) and *Portal 2* (2011), both the antagonist GLaDOS and the protagonist Chell are female. Pictured is concept art of Chell from Valve's headquarters in Bellevue, US. (Claire Bateman)

Respectful representations of women don't necessarily require them to be the 'good guys', as illustrated by the admirably well-characterised female villains in games as diverse as 1985's *Where in the World is Carmen Sandiego?* and 2007's *Portal*.

As hinted above regarding homosexual characters, women are not the only under-represented (or frequently misrepresented) group in video games. And, perhaps not surprisingly, when intersectionality is considered, video games have tended to draw a complete blank. For example, thoughtful representations of non-white female characters – such as the Haitian character Cassandra from 2016's *Mafia III* or the Iñupiaq girl in 2014's *Never Alone* – are extremely rare.

This widespread sexism in video games (however unintentional it may be) has had a cumulative toxic effect, which has at times manifested explosively. In 2012, feminist critic Anita Sarkeesian launched a Kickstarter campaign to crowdfund a video series about the portrayal of female characters in video games, which triggered a campaign of misogynistic harassment against her, including rape and death threats, hacking of her web pages and social media, and online publication of her address and other private details (a practise known as doxing) alongside incitements to violence. Her Wikipedia article was vandalised with racial slurs and sexual images, and she received images of herself being raped by video game characters.

After Sarkeesian started publishing her influential video series, 'Tropes vs. Women in Video Games' (from which much of the material in this chapter is sourced), the attacks escalated. In March 2014, organisers of the Game Developers Choice Awards, at which Sarkeesian was scheduled to speak and receive an award, received an anonymous bomb threat. In August 2014 the increased volume and specificity of the harassment prompted her to leave her home. In October 2014, Sarkeesian cancelled a planned lecture at Utah State University after terrorist

Cassandra from *Mafia III* (2016). *(Hangar 13 / 2K Games)*

Nuna and her Arctic fox from *Never Alone* (2014). *(Upper One Games / E-Line Media)*

Anita Sarkeesian in 2011. *(Anita Sarkeesian / CC BY-SA 2.0)*

threats were received, including one by an individual who claimed to be affiliated with the 'Gamergate' controversy.

Gamergate was essentially a loosely organised anonymous right-wing backlash against progressivism in video games. It started in August 2014 as a personal attack against video game developer Zoë Quinn after her ex-boyfriend published a blog post that erroneously implied she had slept with a video game journalist in exchange for favourable coverage of her interactive fiction game *Depression Quest*. Despite the blog post later being updated to correct the error, Quinn and her family received such extreme levels of misogynistic abuse, along with doxing and threats, that she fled her home for fear of her safety on at least two occasions. This sustained attack grew into a systematic harassment campaign against female video game developers and critics, dubbed Gamergate after actor Adam Baldwin coined the term on Twitter. Attempts to bring legal action against the perpetrators of these attacks and crimes have been unsuccessful, in part because it is difficult to trace which individuals are responsible.

Zoë Quinn in 2016. (Official GDC / CC BY 2.0)

By way of a silver lining, the severity of these attacks has helped to bring issues of progressivism, representation, and the treatment of women in video gaming into the public eye, and has been a catalyst for change. The industry has been gradually responding by improving the way that women and minorities are represented in video games, and by taking steps to prevent harassment in online video game communities and social media platforms.

It may not come as a surprise that video games are *made* predominantly by men, although this situation is slowly improving. In the years from 2005 to 2015, the proportion of female video game professionals worldwide doubled from 11.5 per cent to 22 per cent (although the proportion of females in core content creation roles is much lower). Some countries are lagging behind; for example in 2009, just 4 per cent of game developers in the UK were female.

Meanwhile, female video game streamers and competitors in online tournaments are still routinely subject to shockingly misogynistic abuse. Women and minorities are still woefully under-represented and misrepresented in video games, and in the industry's workforce. There's a long way to go yet.

A 2016 study of 571 games released between 1984 and 2014 found that the sexualisation of female characters was at its height between 1990 and 2005, and then began to significantly decline. (Andre Hunter on Unsplash)

Level 6
CULTURE AND COMMUNITY

Video games have had a profound impact on our culture. Having made their start in amusement arcades in the 1970s, places that were previously associated with gambling and the mafia, video games were treated with suspicion by the moral guardians of the day – and that suspicion has never entirely gone away. Over the years, video games have been blamed for social ills ranging from delinquency to obesity; from rape to school shootings.

The first widespread moral outrage triggered by a video game was in 1976. The 1975 film *Death Race 2000*, starring David Carradine and Sylvester Stallone, depicted a race across the US in which drivers were awarded points for killing pedestrians as brutally as possible. A year later, Exidy released the *Death Race* arcade game, in which the player drove a car over humanoid 'gremlins'. This was the first video game to encourage mass-killing of human-shaped targets, and middle-class America was appalled. The game was criticised as sick and morbid by the National Safety Council, condemned in the *New York Times* and the *National Enquirer*, and hotly discussed on popular TV shows such as *60 Minutes*. Ironically, the controversy led to a large jump in sales. Game publishers have been courting controversy ever since.

By 1979, a middle-class mother from Long Island, New York called Ronnie Lamm was at the head of a passionate campaign against video games, leading to some American towns banning arcades altogether. In an interview with the *New York Times*, Lamm said of arcade games, 'They are not wholesome. They mesmerize our children; they addict them and force them to mindlessly pour one quarter after another into the slots.' But protests against specific games continued to have the unintended effect of making those games more popular. In 1982, a producer of X-rated videotapes branched out into video games with *Custer's Revenge*, in which the player controlled an unclothed General Custer with an oversized erection, trying to cross the screen and rape a bound Native American woman. Organisations such as Women Against Pornography and the American Indian Community House protested against the game and pressured legislators to ban it. *Custer's Revenge* ended up selling 80,000 copies, twice as many as the developer's other adult titles. The game is a regular feature on lists of the worst games ever made.

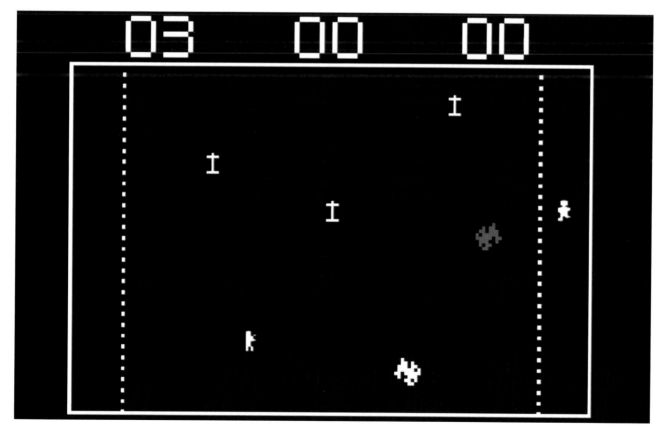

Death Race (1976). *(Exidy)*

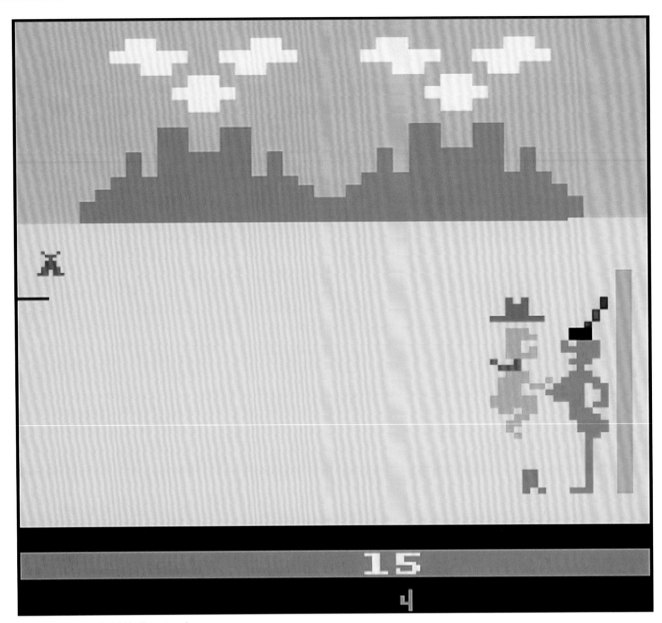
Custer's Revenge (1982). (Mystique)

But the biggest moral panic was yet to come. In the early 1990s, cutting-edge games were becoming more photorealistic, featuring digitisations of real actors; and 1993's *Mortal Kombat* applied this technology to the one-versus-one martial arts fighting genre made popular by *Street Fighter II*. *Mortal Kombat* allowed victorious players to perform gruesomely graphic 'fatalities' on their opponents, which triggered a flood of controversy and spurred United States Senator Joe Lieberman to initiate a widely publicised joint Congressional hearing on violence in video games. Lieberman accused video games of inspiring real-life violence and argued that 'instead of enriching a child's mind, these games teach a child to enjoy inflicting torture.' As well as *Mortal Kombat*, Lieberman and his colleagues singled out *Night Trap*, a previously obscure game in which the player tries to rescue women from vampires – though many of the panellists clearly hadn't played *Night Trap* themselves since they incorrectly claimed the goal of the game was to kill women. Ultimately, a bill was introduced threatening to impose a government-run ratings system upon the industry. The Entertainment Software Association responded by creating the self-regulatory ratings system that is still in use across the US today; other countries followed suit.

Academic studies conflict on whether there are links between video game violence and real-world aggression. Regardless of the extent of such a link, video games have often been used as a scapegoat for wider societal problems, particularly in the US. In 1999, one of the perpetrators of the Columbine High School massacre in which 13 people were murdered said, when planning the shooting, 'It's going to be like fucking *Doom*,' referring to the nihilistically violent 1993 video game. Video games were blamed for the shooting, prompting lawmakers to try and prohibit violence in video games. (The shooting itself inspired a video game called *Super Columbine Massacre RPG!*, intended as a satirical commentary on how traditional media sensationalised the crime, although it was received badly.) This pattern of blaming video games was repeated after the 2012 Sandy Hook Elementary School shooting in which 26 children and teachers were murdered – despite the perpetrator's most played video games reportedly being the decidedly non-violent *Dance Dance Revolution* and *Super Mario Bros. 3*.

A 'fatality' from Mortal Kombat (1992). (Midway)

Night Trap (1992). (Digital Pictures / Sega)

Ironically, the head of the National Rifle Association denounced video game makers as 'a callous, corrupt, and corrupting shadow industry that sells and sows violence against its own people.' And again, after the 2018 Parkland school shooting which claimed 17 young lives, President Donald Trump blamed video games, saying the 'level of violence on video games is really shaping young people's thoughts'.

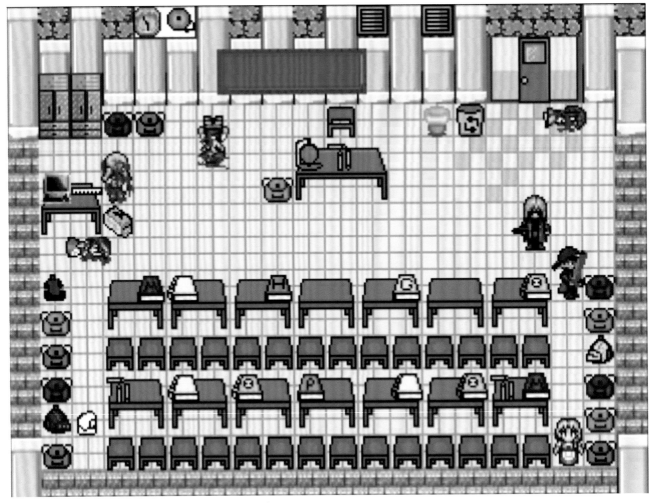

Super Columbine Massacre RPG! (2005). *(Danny Ledonne)*

The video game franchise that has arguably garnered the most controversy is also one of the best selling of all time: *Grand Theft Auto*, developed by British company Rockstar North (formerly DMA Design). Starting with the first instalment of the gangster-themed series, which appeared in 1997, *Grand Theft Auto* drew condemnation for glorifying carjackings, bank robberies, and casual homicide. 2001's *Grand Theft Auto III* allowed players to have sex with prostitutes and then murder them to reclaim the payment, with no significant consequences. Yet the biggest public outrage surrounding the franchise wasn't about the violence; instead, it centred around a minigame from 2004's *Grand Theft Auto: San Andreas* that depicted the player's character having consensual (and mostly clothed) sex with his in-game girlfriend. The minigame didn't even appear in the final release, but a Dutch software modder found the hidden code and published the scene, sparking a media frenzy, lawsuits, and US congressional debate. Despite the rest of the game featuring considerably more morally questionable content (and perhaps because parents had been unaware of that content before the game hit the headlines), the public reaction was outrageous enough to trigger several documentaries including a BBC movie and book. The panic did nothing to stall the fortunes of the franchise – in 2013, sales of *Grand Theft Auto V* passed the $1 billion mark just three days after launch.

As much as video games have been demonised, there has also been a growing recognition that they can be a positive part of society. Research has shown that video gaming can improve emotional well being, develop cognitive skills, and encourage friendships. Some institutions have recognised video games as works of art, and important artefacts of our culture that are worth preserving.

Grand Theft Auto: San Andreas (2004) 'Hot Coffee' mod. *(Rockstar North / Rockstar Games)*

Museum of the Moving Image, New York. (NickCPrior / CC BY-SA 3.0)

As early as 1989, the Museum of the Moving Image in New York displayed a retrospective of video games (and has maintained the display ever since). The curator of that exhibit, Rochelle Slovin, was one of the first advocates of video games as art: 'The stripped-down feel of the early games suggests early black-and-white films before sound. Here, too, the art is defined by its limitations. There is a poetic spareness about the early moments in each technology.'

Since then, many museums and art institutions have followed suit. The Library of Congress in Washington DC has been collecting and preserving video games and related artefacts since 1998. In 2002, the Barbican Centre launched its major 'Game On' exhibit exploring the history and culture of computer games, which has toured to more than 20 countries and been seen by over 2 million people. By 2012, the link between video games and art was mainstream, with such venerable institutions as New York's Museum of Modern Art and the Australian Centre for the Moving Image curating collections.

But the idea of video games as art is controversial. Film critic Roger Ebert said that 'video games can never be art' because they are meant to be won rather than experienced, and because 'No one in or out of the field has ever been able to cite a game worthy of comparison with the great filmmakers, novelists and poets.'

Barbican International Enterprises' interactive touring exhibition 'Game On' featured over 150 playable video games. Pictured here at the Swedish National Museum of Science and Technology in 2014. (/kallu / CC BY-SA 2.0)

Whether or not games are art, they have inspired many artists and entertainers in other media. As video game graphics improved, players started creating stories around recordings of gameplay. In 1996, a 100-second short film called *Diary of a Camper*, which used footage of the first-person shooter *Quake* to tell its story, inspired a whole community of gamer artists. The form of using video game graphics engines to create cinematic productions became known as 'machinima', a term coined in 2000 by machinima maker and distributor Hugh Hancock.

A scene from *Diary of a Camper* (1996). *(United Ranger Films / id Software / GT Interactive)*

Some artists have used art to explore and subvert the experience of playing video games, such as the European art collective JODI. Their artistic modifications of video games serve as commentary on gaming and internet culture; for example 2013's *SOD* is a mod of early first-person shooter *Wolfenstein 3D* that reduces the graphics to abstract black and white shapes.

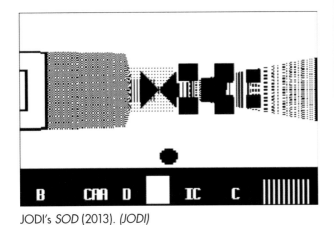

JODI's *SOD* (2013). *(JODI)*

Still from the *Q*Bert* cartoon, part of the Saturday Supercade animated television series. *(Ruby-Spears Enterprises / Worldvision Enterprises)*

In 1982, American sketch comedy show *Saturday Night Live* introduced a character called Alan the Video Junkie, satirising the addictiveness of video games. In 1983, American reality TV show *That's Incredible!* aired the first televised video game competition. The first TV series centred around competitive video gaming was the UK's *Gamesmaster*, which aired from 1992 to 1998. There have even been entire TV channels dedicated to video games, such as North America's G4, which aired from 2002–2014.

The first feature film based on a video game was 1993's *Super Mario Bros.* starring Bob Hoskins and John Leguizamo as Mario and his brother Luigi. It was a critical and commercial failure, starting a long-standing Hollywood tradition of turning great games into terrible films. Games to get the treatment included *Double Dragon, Street Fighter, Mortal Kombat, Wing Commander, Final Fantasy, Lara Croft: Tomb Raider, Resident Evil, House of the Dead, Alone*

Many early video games were derivative of other forms of media such as films, television and books – but gradually this became a two-way relationship. In the early 1980s there were TV cartoon shows based on arcade games *Pac-Man, Pole Position* and *Q*Bert*. By the end of the 1980s Nintendo was promoting its games with TV series *The Super Mario Bros. Super Show!* and *The Legend of Zelda*. Nintendo's 1996 *Pokémon* game spawned a TV series (which has been running continuously since 1997), a collectible card game (ditto), and two dozen feature films.

Video games have long been a staple of television. As far back as 1978, *TV POWWW* was a gameshow in which home viewers controlled a video game via telephone.

Cards from the Pokémon collectible card game. In 2019, a single Pokémon card was sold at auction for $195,000. *(Minh Hoang / CC BY-ND 2.0)*

Astronomer Patrick Moore as the GamesMaster. *(Hewland International / Channel 4)*

A film poster for *Pokémon Detective Pikachu*, in Newport, Wales. *(Jaggery / CC BY-SA 2.0)*

in the Dark, Doom, BloodRayne, Silent Hill, DOA: Dead or Alive, Postal, Hitman, Max Payne, Tekken, Prince of Persia: The Sands of Time, Need for Speed, Ratchet & Clank, Warcraft*, and *Assassin's Creed* – none of which received an aggregated review score of higher than 46 per cent from critics, according to reviews website Rotten Tomatoes. Two 2019 films finally bucked the trend, with *Pokémon Detective Pikachu* and *The Angry Birds Movie 2*, which were mostly positively received by critics, scoring a heady 68 per cent and 77 per cent respectively on the Tomatometer.

However, films inspired by video gaming in general (rather than a specific game) have fared better. Much-loved examples include 1982's *Tron*, Disney's prototypical sucked-into-a-video-game movie which broke new ground with its ultra-cool ultra-neon portrayal of the digital landscape of gaming. The 1983 film *WarGames* played on Cold War paranoia, telling the story of a teen hacker who believes he is playing a computer game but is inadvertently manipulating real world leaders towards global thermonuclear war. The 1989 movie *The Wizard* told the story of a traumatised boy who attempts to escape his demons by entering a video game tournament – the film showed video gaming in a very positive light (and featured a lot of Nintendo product placement). In 2012, animated Disney movie *Wreck-It-Ralph* dove deep into arcade video gaming nostalgia, featuring a cast of 1980s video game characters; it garnered an Oscar nomination for Best Animated Feature, as did its 2018 sequel. 2018 also saw Steven Spielberg's blockbuster film *Ready Player One* about a worldwide virtual reality game, based on the Ernest Cline

Promoting the first Angry Birds movie in 2016, at the Brussels Comic Con. *(Miguel Discart from Bruxelles, Belgique / CC BY-SA)*

Still from *The Wizard* (1989). *(BagoGames / CC BY 2.0)*

Characters from *Wreck-It Ralph* (2012). *(Jared / CC BY 2.0)*

Still from *Tron* (1982). *(Walt Disney Productions)*

novel of the same name, and dripping with video game nostalgia.

There have, of course, been many other novels about video games. The plot of Terry Pratchett's 1992 novel *Only You Can Save Mankind* revolved around a 12-year-old boy helping the alien invaders from a fictional video game return home. Kit Reed's 2000 novel *@Expectations* tells the story of a woman who escapes her boring existence by inhabiting a virtual world. Jason Rekulak's 2017 novel *The Impossible Fortress*, about a 14-year-old determined to make his own video game, literally includes a video game – the first few lines of each chapter are BASIC programming code that can be put together to make a playable game.

Video games have also featured in short stories, such as David K. Smith's *Above Candles*; and graphic novels, such as Cory Doctorow's *In Real Life*.

As well as books, there have been many much-loved magazines about video games. The first were *Play Meter*, which ran from 1974 to 2018, and *RePlay*, which started in 1975 and continues online to this day (electronic only since 2017). The 1990s were a golden age for video game magazines; many playful and colourful titles started their runs between 1990 and 1993, including *Mean Machines*, *Mega*, *Total!*, *Super Play*, *Nintendo Magazine System* (later *Official Nintendo Magazine*), *GamesMaster*, and *Edge*. As of 2019 the only one still in print (as well as online) is *Edge*.

Flyer showing two early video game magazines. *(The International Arcade Museum)*

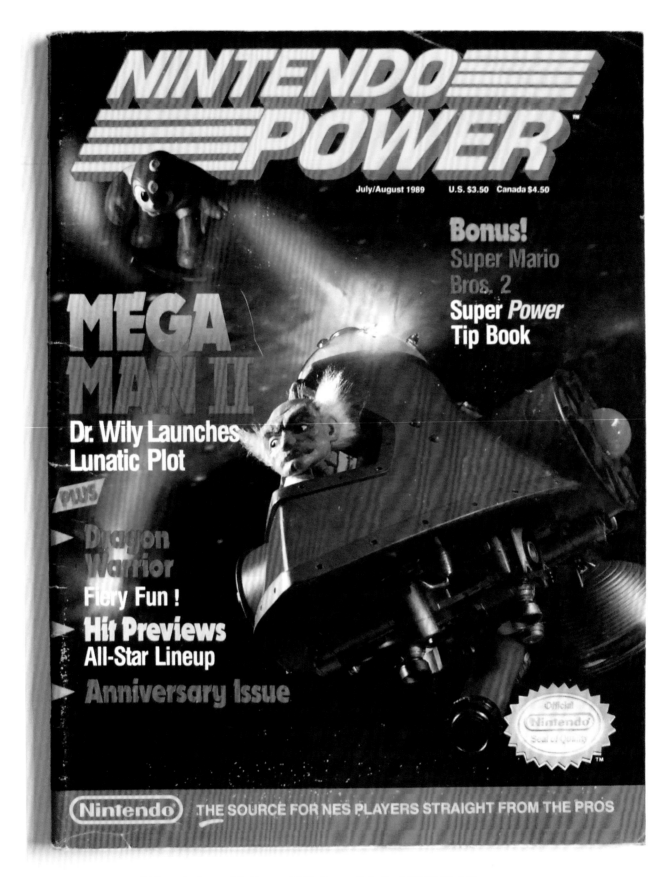

First anniversary issue of *Nintendo Power*, July/August 1989. *(Bryan Ochalla / CC BY-SA 2.0)*

But the media genre that has arguably been most profoundly influenced by video games is music. The first commercially available video games, like *Computer Space* and *Pong*, featured monophonic beeps. By the end of the 1970s, sound started to be used more creatively – *Space Invaders* had a simple four-tone tune that sped up as the action escalated, ramping up the tension surprisingly effectively. Bands such as Yellow Magic Orchestra were inspired by the synthesised electronic music used in video games, which grew gradually more sophisticated throughout the 1980s. In 1986, the Golden Joystick Awards (one of the longest running video game awards ceremonies)

The first issue of *Edge*, October 1993. *(Ian Dick / CC BY 2.0)*

introduced a category for Best Soundtrack of the Year; and around that time game soundtracks started being released on cassette in Japan, encouraging game makers to take their soundtracks more seriously. Starting with the music of the *Final Fantasy* series, composed by Nobuo Uematsu, video game music took on film score quality, complete with full orchestral and vocal tracks. This attracted well-known musicians to compose music for video games, such as Trent Reznor and Nine Inch Nails' *Quake* score. Video game music has influenced mainstream music, particularly hip

Yellow Magic Orchestra in 2008. *(The_Junes of Flickr.com / CC BY 2.0)*

Journey (1983). *(Bally Midway)*

hop, pop and electronica – and has in some cases become mainstream in its own right: in 2012 a song from *Civilization IV* won a Grammy; and there is a branch of academia called ludomusicology, dedicated to studying video game music.

As the video game market has grown, licensing existing songs for use in video games has become a lucrative opportunity for rights holders. One of the first games to use licensed music was Midway's 1983 arcade game *Journey*, which featured 8-bit arrangements of the titular band's music. Music-themed games such as *Dance Dance Revolution*, *SingStar*, *Guitar Hero* and *Just Dance* also serve as a showcase for licensed music, releasing new versions regularly to give gamers access to the latest popular songs.

Another area of culture which video games have profoundly impacted is education. Ever since 1985's *Where in the World is Carmen Sandiego?* taught geography by having the player chase the eponymous thief across the globe, video games have been used to teach children every subject on the curriculum. Jump forward to the present day, and video games are a ubiquitous ingredient in almost every child's

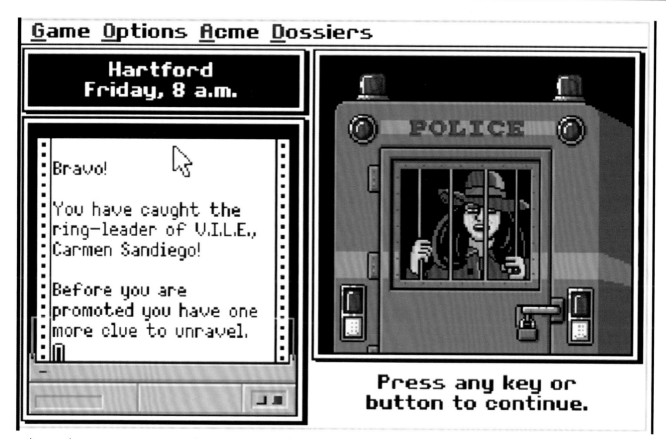

Where in the U.S.A. is Carmen Sandiego? (1986). (Broderbund)

education; and Carmen Sandiego stars in her own Netflix animated series. Video games are also used to educate adults, such as simulations used in the military to teach tactical skills.

The video gaming community itself has a rich culture. The earliest video games, in the 1960s, were the preserve of academics and researchers with access to expensive mainframe computer equipment. The first commercial video games were in bars and nightclubs, therefore played only by adults. When video games hit the arcades in the early 1970s, and became immensely popular, they became accessible to teenagers as well. Although most games were single-player, the public space meant that gaming was an inherently social activity. Arcades attracted a wide range of people, including those who were otherwise marginalised – academic Dmitri Williams explains that 'for the awkward, the underclass, or the socially restricted player, success at a game translated into a level of respect and admiration previously unavailable outside of the arcade. There was no gender or status bias in arcade competition, and the machine didn't care if the player was popular, rich or an outcast.'

In the 1970s, when video games entered the home, they were marketed as a family activity, and increasingly appealed to younger children. As those children grew up, and games explored more mature themes, the audience broadened, and has continued to do so. As of 2019, the average age of video gamers worldwide was 35, which is also approximately the average age of every human being alive today.

Many video games are played solo, but there is a long tradition of multiplayer gameplay, and gaming communities exist in many forms. The first significant in-game community formed within a text-based computer game created by British students Roy Trubshaw and Richard Bartle, called *MUD* – short for *Multi-User Dungeon*. The game could be played by a large number of players simultaneously, connected over the ARPANET (the predecessor of the internet). *MUD* offered the player a chance to adopt a new persona and explore a virtual fantasy world, socialising with other players in real-time. It was distributed for free, and widely adapted throughout the 1980s. Its cultural impact on the gaming community included coining the term 'newbie' or 'noob' to describe a player new to the game; and normalising gender-swapping, with male players often inhabiting female online personas and vice-versa.

In 1986, American company LucasArts attempted to take the idea of large-scale online gaming to the next level by creating a graphical virtual world. Their game, *Habitat*, is considered the forerunner of modern Massively Multiplayer Online Role-Playing Games (MMORPG, a term coined by Richard Garriott, creator of the *Ultima* series of games). Ultimately, *Habitat* was deemed not to be economically viable and never made it past beta.

Yet MMORPGs have grown in popularity ever since, inhabited by so many players that their virtual worlds have real economies, with players creating and selling in-game items for real-world money. The biggest MMORPG to date is *World of Warcraft*, released in 2004, which at its peak had over 11 million subscribers, and an in-game economy with the Gross Domestic Product of a small country.

Another common way for large numbers of people to play a video game simultaneously is using computers

```
Telnet british-legends.com

Initialised.

Multi-User Dungeon - MUD1 Version 3E(19)

             You are invited to check out Section 9,
             our discussion forum for MUD players.

                    Please direct your browser to:
             http://www.british-legends.com/Forums/S9.htm

             *******************************

             ***************************************************
             * MUD2.COM is where you'll find the next generation *
             * version of MUD1/British Legends. Another creation *
             *      of Richard Bartle, MUD2 offers many extras,  *
             *    including smart mobiles, new areas, and more.  *
             *    Best of all, it's free. Why not try it today?  *
             ***************************************************

Origin of version: Fri Jan 19 22:26:12 2018

Welcome! By what name shall I call you?
*
```

MUD1 (1978). *(Richard Bartle)*

Habitat (Beta) (1986). *(Lucasfilm Games / Quantum Link / Fujitsu)*

World of Warcraft (2004). *(Blizzard Entertainment)*

DreamHack LAN party, 2004. *(Toffelginkgo / CC BY-SA 3.0)*

linked by a Local Area Network (LAN). So-called LAN parties, popularised in the mid-1990s by games such as the gory first-person shooter *Quake*, often involve each player bringing their own computer – the computers are then linked together using ethernet cable. Playing games over a LAN has a social advantage over playing games over the internet, because players can see and talk to each other. The biggest LAN party in the world is at the

World Cyber Games, 2004. (Peter Kaminski from San Francisco, California, USA / CC BY 2.0)

annual DreamHack 'digital festival' in Sweden, which physically connects over 20,000 computers.

These technologies allowed for the growth of competitive video gaming tournaments, which in turn has led to the rise of professional video gaming. Following the 1997 Asian financial crisis, unemployment in South Korea was high. To help stimulate the economy, the Korean government made a huge investment in building broadband networks, and out-of-work citizens increasingly spent their time playing video games in internet cafés known as PC Bangs. Competitive video gaming became a national pastime, to the extent that the government founded the Korean e-Sports Association in 2000 to promote and regulate esports in the country. The Korean e-Sports Association founded the World Cyber Games, modelled after the Olympic Games, which remains one of the largest global esports tournaments.

The scale of these tournaments has grown to dizzying proportions. In *Fortnite*, released in 2017, up to 100 players are dropped into a virtual world with the aim of killing all the other players and being the last person standing. The 2019 *Fortnite* World Cup online competition was entered by 40 million players. The total prize fund for the tournament was over $30 million. The final of the competition, with the best 100 players, happened in front of a live audience in the Arthur Ashe Stadium in New York (capacity 23,000 people, although 2 million more watched the event live online).

Fortnite (2017). (Public domain)

The tournament made international news when 16-year-old Kyle Giersdorf walked away with the $3 million first prize – that's more than Novak Djokovic won for his victory at the 2019 Wimbledon Tennis Championship.

The designation of competitive video gaming as a 'sport' is controversial, but is gradually becoming more widely accepted – to the extent that the International Olympic Committee (IOC) has considered including video gaming events as part of the Olympics. From a 2019 IOC press release: 'Many sports simulations are becoming more and more physical thanks to Virtual and Augmented Reality which replicate the traditional sports. The International Federations are encouraged to consider how to govern electronic and virtual forms of their sport and explore opportunities with game publishers.'

Watching people play video games is not restricted to major international tournaments. In 2011, a video streaming service called Twitch was launched, primarily to allow people to broadcast themselves playing video games. Within two years, the service had 45 million users. By 2018, it featured over 2 million broadcasters and 15 million daily active users. Most of this content is consumed for free, but broadcasters can make an income from advertising revenue, tips, and paid subscriptions in return for additional features.

After the 1983 video game crash nearly wiped out the budding global video game industry (see Level 2: Platforms and Technology), many media outlets speculated that video games had been a fad, on a par with hula hoops. They couldn't have been more wrong. Video games have become a huge part of our global culture, and they're here to stay.

Final Fantasy VI (1994) is one of the best-loved instalments of the Final Fantasy role-playing game franchise. *(Square)*

Level 7
GAMES

It's beyond the scope of this book to delve into detail about the many thousands of notable video games, but this chapter lists some that have made the biggest impact.

TOP TEN BEST SELLING ARCADE GAMES OF ALL TIME

These are the arcade video games that have sold the most units, but they are not necessarily the arcade games that have raked in the most coins. If you adjust for inflation, there are arcade video games that have grossed more money than some on this list – the most prominent contenders being the bloody fighting game *Mortal Kombat* from 1991, and the two-on-two basketball game *NBA Jam* from 1993.

But for sheer numbers of cabinets sold, these are the champions.

Sales figures are estimated, and hard to verify. Some of the figures below relate to US sales only. In terms of ranking, the top nine are undisputed, but sources vary on which game is in the number ten slot. The candidates, all of which are estimated to have sold over 30,000 units, are: *Donkey Kong Jr.* (1982) in which the giant ape Donkey Kong has been kidnapped by Mario, and the player must swing through

Mortal Kombat (1992). *(Scott Schiller / CC BY 2.0)*

NBA Jam (1993). *(Midway)*

Donkey Kong Jr. (1982). (Marco Verch / CC BY 2.0)

Out Run (1986). (Tiia Monto / CC BY-SA 3.0)

the jungle to rescue him; *Mr. Do!* (1982), which featured a circus clown (or a snowman in the Japanese version) digging for cherries; and *Out Run* (1986), where you could climb into the cockpit-style cabinet to race in a Ferrari Testarossa Spider. In addition, there are Japanese gambling video games such as horse race game *Starhorse* that, if you combine the variant versions, have sold more than 30,000 units.

The rest of the top-selling arcade video games follow.

Number nine: *Galaxian* (1979)
Cabinets sold: 40,000
Created to compete with the phenomenally successful *Space Invaders*, and with very similar gameplay (shooting down invading alien fleets), this game boasted more colours and harder enemies.

Galaxian (1979). (Jordiferrer / CC BY-SA 3.0)

Number eight: *Centipede* (1981)
Cabinets sold: 56,000
Similar to *Galaxian*, in this game the player was at the bottom of the screen shooting at enemies descending from above, but instead of aliens – insects!

Centipede (1981). (joho345 / CC BY-SA 3.0)

Defender (1981). *(Rob DiCaterino / CC BY 2.0)*

Number seven: *Defender* (1981)
Cabinets sold: 60,000
Shooting aliens again, but this time in an innovative side-scrolling format which allowed the player to explore beyond the boundaries of a single static screen.

Number six: *Asteroids* (1979)
Cabinets sold: 100,000
The player controlled a space ship being bombarded by asteroids on all sides – which, when shot, broke into two. Controlling the ship was tricky because it kept its momentum.

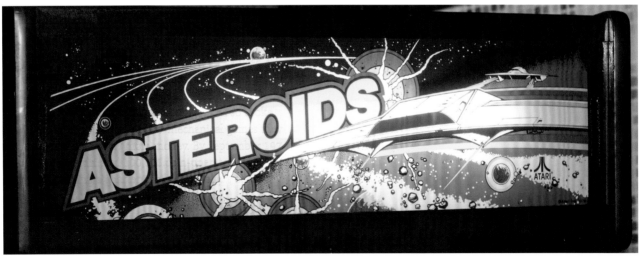

Asteroids (1979). *(Steven Miller / CC BY 2.0)*

Number five: *Ms. Pac-Man* (1981)
Cabinets sold: 115,000
The first playable female protagonist in a video game. Featured similar gameplay to its older brother *Pac-Man* (being chased by ghosts in a maze), but with new maze designs and more randomised enemies.

Number four: *Donkey Kong* (1981)
Cabinets sold: 132,000
The player climbed scaffolding to save a woman from an angry giant gorilla. Introduced a new genre – platforming – and more story-driven gameplay. This is the game that catapulted Nintendo into the American market.

Ms. Pac-Man (1982) and *Donkey Kong* (1981). *(Rob Boudon from New York City, USA / CC BY 2.0)*

Number three: *Street Fighter II* (1991)
Cabinets sold: 200,000
Everyone assumed video game arcades were in terminal decline, until this cartoonish one-versus-one fighting game took the world by storm. Three decades after its release, this game is still a regular feature in the competitive scene.

Street Fighter II (1991). *(Jonathan Sloan from Burnaby, Canada / CC BY 2.0)*

Number two: *Space Invaders* (1978)
Cabinets sold: 360,000
Launched a golden age of video games, and spawned countless imitators. This game, with its ever-descending alien hordes, has become the archetype of video games – to the extent that this book signals that it's about video games by including pixelated aliens on the cover.

Space Invaders (1978). *(Jordiferrer / CC BY-SA 3.0)*

Number one: *Pac-Man* (1980)
Cabinets sold: 400,000
Most video games were simple sports games or space shooters – until *Pac-Man*. Its phenomenal success was down to its addictive maze-chase gameplay, but also because it appealed to a broad demographic, including women. The character, whose shape was inspired by a pizza with a slice missing, is almost as widely recognised as Mickey Mouse.

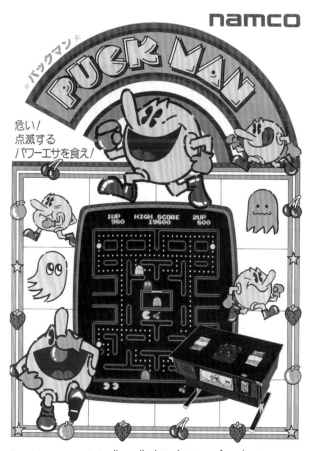

Pac-Man was originally called Puck Man after the Japanese word for flapping, ぱくぱく (pakupaku), but there were concerns about the arcade cabinets in English-speaking countries being defaced. *(The International Arcade Museum)*

TOP TEN BEST SELLING CONSOLE GAMES OF THE 1980S

After the collapse of Atari catalysed a global crash in the video game market, Nintendo swooped in and rose phoenix-like, building up to almost total domination in the 1980s. This ubiquity firmly fixed their franchise characters – especially the gymnastic Italian plumber Mario – in the popular consciousness.

The number ten slot on this list is disputed. *Pitfall!* is a side-scrolling Indiana Jones-inspired platformer released in 1982 for the Atari 2600, which sold in excess of 4 million units, although the exact figure is unknown.

Again, in general, sales figures are hard to verify; the figures in the following sections are approximate.

Pitfall! (1982). (Digital Game Museum / CC BY 2.0)

Number ten: *Excitebike* (1984)
Units sold: 4.2 million
Platform: Nintendo Entertainment System (NES)
This motocross racing game may have faded into relative obscurity today, but it was one of the NES's earliest megahits, with its thrilling gameplay and ability to create your own tracks.

Excitebike (1984) NES cartridge. (Chris Harrison / CC BY 2.0)

Number nine: *Zelda II: The Adventure of Link* (1987)
Units sold: 4.4 million
Platform: Nintendo Entertainment System (NES)
A role-playing game in which the player explored a large map and entered dangerous dungeons as part of a quest to save Princess Zelda. The success of this sequel meant there would be many more Zelda games to come.

Zelda II: The Adventure of Link (1987). (Nintendo)

Number eight: *The Legend of Zelda* (1984)
Units sold: 6.5 million
Platform: Nintendo Entertainment System (NES)
The heroic fantasy game that launched a franchise, *The Legend of Zelda* often appears on lists of the most influential games of all time. Princess Zelda was named after American novelist and socialite Zelda Fitzgerald.

Link from *The Legend of Zelda* (1984). (Public domain)

Number seven: *Pac-Man* (1982)
Units sold: 7.0 million
Platform: Atari 2600
The technological capability of the Atari 2600 home console was not advanced enough to preserve the full *Pac-Man* arcade experience, but that didn't stop it selling by the truckload.

Number six: *Super Mario Bros. 2* (1984)
Units sold: 7.5 million
Platform: Nintendo Entertainment System (NES)
After the incredible success of the first *Super Mario Bros.* game, Nintendo needed a sequel. The Japanese sequel was deemed too similar to the original, and too difficult for overseas players, so a promotional game made for Fuji Television called *Doki Doki Panic!* was reskinned and released as *Super Mario Bros. 2*. Which explains why this platformer is so different to every other Mario game.

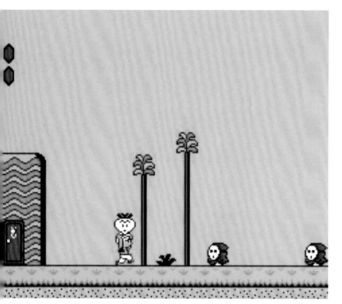

Yume Kōjō: Doki Doki Panic (1987). *(Nintendo)*

Number five: *Super Mario Bros. 3* (1984)
Units sold: 17.3 million
Platform: Nintendo Entertainment System (NES)
The gameplay of the Mario series is timelessly satisfying – jumping on walking mushrooms and turtles on a mission to save the princess from the spiny-shelled Bowser. This instalment introduced new power-ups, including one that allowed Mario to fly. *Super Mario Bros. 3* has aged well, and regularly appears on lists of the best games of all time.

Super Mario Bros. 3 (1984). *(Martin Bergesen / CC BY 2.0)*

Number four: *Super Mario Land* (1989)
Units sold: 18.4 million
Platform: Game Boy
Nintendo's handheld Game Boy would go on to be the best selling console in any format, until 2007 when it was overtaken by the PlayStation 2. So it's no surprise that, despite being monochromatic, Mario's first outing on the Game Boy did well.

Number three: *Duck Hunt* (1984)
Units sold: 28.3 million
Platform: Nintendo Entertainment System (NES)
Came with a light gun so you could shoot ducks on the screen. Its impressive sales are due to the fact it was bundled with the NES.

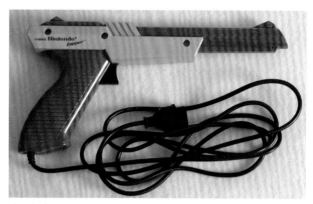

The light gun accessory used to play *Duck Hunt* (1984). *(methodshop.com / CC BY-SA 2.0)*

Number two: *Tetris* (1984)
Units sold: 30.3 million
Platform: Multi-platform
An addictively simple puzzle game of falling shapes. The Nintendo version was bundled with the Game Boy, and familiarised millions of gamers worldwide with the tune of the Russian folk song *Korobeiniki*.

Game Boy with *Tetris* (1989). *(Sammlung der Medien und Wissenschaft / CC BY 4.0)*

Number one: *Super Mario Bros.* (1985)
Units sold: 40.2 million
Platform: Nintendo Entertainment System (NES)
The game that is synonymous with Nintendo. The game has been praised for the finesse with which the player can control Mario. Its levels were designed to gently teach the core concepts of the game without having to resort to an instruction manual.

A combined *Super Mario Bros.* (1985) and *Duck Hunt* (1984) cartridge in a Nintendo Entertainment System. *(Yagamichega / CC BY 3.0)*

TOP TEN BEST SELLING CONSOLE GAMES OF THE 1990S

The 1990s saw Nintendo's stranglehold on the video games market challenged by Sega and Sony, but Nintendo ultimately came out on top, bagging seven of the top ten slots.

Number ten: *Super Mario All-Stars* (1993)
Units sold: 10.6 million
Platform: Super Nintendo Entertainment System (SNES)
Not a new game, but the first four games of the Mario series collected onto one cartridge. Many hours of joy.

Number nine: *Gran Turismo* (1997)
Units sold: 10.85 million
Platform: PlayStation
An exhilarating racing game that pushed the boundaries of realism, with a killer soundtrack.

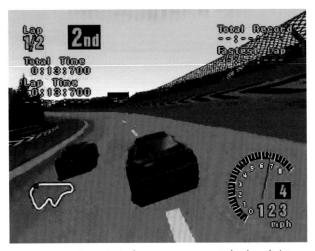

Gran Turismo (1997). (Polys Entertainment Cyberhead / Sony Computer Entertainment)

Number eight: *Super Mario Land 2: 6 Golden Coins* (1992)
Units sold: 11.2 million
Platform: Game Boy
Introduced Mario's evil doppelganger Wario. This game didn't break any new ground, but that didn't stop it from being great fun.

Super Mario 64 (1996). (Nintendo)

Number seven: *Super Mario 64* (1996)
Units sold: 11.9 million
Platform: Nintendo 64
Not the first 3D platformer, but its open-world gameplay that rewarded you for exploring made it one of the most popular and enduring.

Number six: *Pokémon Yellow* (1998)
Units sold: 14.6 million
Platform: Game Boy
A role-playing game in which you captured cute little monsters and battled them against each other. This was an enhanced version of the first *Pokémon* game that had been released two years earlier.

Pokémon Yellow (1998). (Adam Purves (S3ISOR) / CC BY 2.0)

Number five: *Sonic the Hedgehog* (1991)
Units sold: 15.0 million
Platform: Sega Mega Drive
Brought speed and attitude to the platforming genre. Sonic, a sassy blue hedgehog, was so cool he briefly overshadowed the monolithic Mario franchise.

Sonic the Hedgehog (1991). (Patrick Janicek / CC BY 2.0)

Number four: *Lemmings* (1991)
Units sold: 20.0 million
Platform: Multi-platform
Originally released on Commodore's Amiga personal computer, this quirky puzzle game featured a stream of suicidal lemmings that the player had to rescue by removing and adding obstacles.

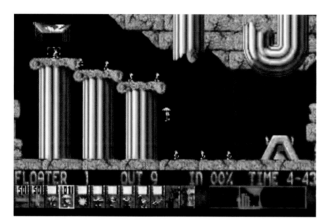

Lemmings (1991). (DMA Design / Psygnosis)

Number three: *Super Mario World* (1990)
Units sold: 20.6 million
Platform: SNES
The peak of family-friendly two-dimensional platforming. Introduced Mario's steadfast companion Yoshi, a green dinosaur-like character that the player could ride.

Number two: *Pokémon Gold* and *Silver* (1999)
Units sold: 23.1 million
Platform: Game Boy Color
In 1998, the full-colour version of the Game Boy came out, and it was only a matter of time before it got its own *Pokémon* game. This sequel improved on the original, offering more strategy and more monsters.

Number one: *Pokémon Red*, *Green* and *Blue* (1996)
Units sold: 31.4 million
Platform: Game Boy
The original *Pokémon* game that gave birth to a multimedia behemoth. The blue (green in Japan) and red versions of the game were the same except each featured an exclusive extra monster to collect.

Pikachu, the Pokémon mascot, has headed up a massive merchandising empire. (Voltordu from Pixabay)

TOP TEN BEST SELLING CONSOLE GAMES OF THE 2000S

Sony may have sold more consoles than Nintendo in the 2000s, but Nintendo still had most of the best selling games.

Number ten: *Nintendogs* (2005)
Units sold: 24.0 million
Platform: Nintendo DS
A pet-raising simulation in which you could play with, train, pet, walk, brush, and wash a virtual dog.

Number nine: *Grand Theft Auto IV* (2008)
Units sold: 25.0 million
Platform: Multi-platform
Debuting on the PlayStation 3, this game followed the gritty story of an immigrant trying to escape his past while under pressure from loan sharks and mob bosses. More serious in tone than others in the series, and considered one of the best.

Number eight: *Grand Theft Auto: San Andreas* (2004)
Units sold: 27.5 million
Platform: Multi-platform
A gangster simulation, but also so much more. A big open world to explore, in which you could fly jets, rob banks, gamble, buy property, drive a train, play basketball… the list goes on.

Nintendogs (2005). (Nintendo)

Number seven: *Wii Play* (2006)
Units sold: 28.0 million
Platform: Nintendo Wii
A collection of simple games showcasing the Wii's motion control interface, including fishing, billiards, and… posing. Bundled with the console.

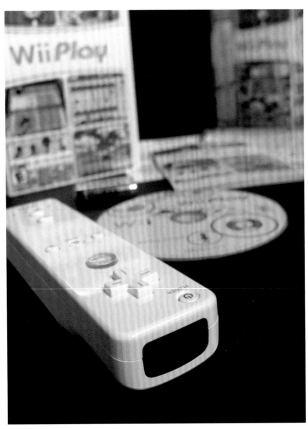

Wii Play (2006). *(Michel Ngilen / CC BY 2.0)*

Number six: *New Super Mario Bros. Wii* (2009)
Units sold: 30.3 million
Platform: Nintendo Wii
Originally released on the handheld Nintendo DS, but the Wii version introduced a simultaneous multiplayer mode.

Number five: *New Super Mario Bros.* (2006)
Units sold: 30.8 million
Platform: Nintendo DS
Classic two-dimensional Mario platforming.

Number four: *Wii Sports Resort* (2006)
Units sold: 33.1 million
Platform: Nintendo Wii
Came with the Wii MotionPlus accessory that allowed for finer motion control. Featured 12 sports including sword fighting, frisbee, and canoeing.

Number three: *Mario Kart Wii* (2008)
Units sold: 37.2 million
Platform: Nintendo Wii
The best selling version of the much loved kart racing series. The game was sometimes packaged with a pair of plastic steering wheels into which the motion controllers could fit, allowing for a more intuitive driving experience.

Number two: *Wii Fit* (2007)
Units sold: 43.8 million
Platform: Nintendo Wii
Sold with a 'balance board' on which the player stood while carrying out a series of gamified exercises. These sales figures include the enhanced version of the game, *Wii Fit Plus*.

Playing on the Wii balance board. *(Lukas Mathis / CC BY-SA 2.0)*

Number one: *Wii Sports* (2006)
Units sold: 82.9 million
Platform: Nintendo Wii
Five minigames – tennis, baseball, bowling, golf and boxing – designed to show off the innovative motion controls of the Wii. Packaged with the Wii, which is one of the best selling consoles of all time.

Playing Wii Sports (2006). *(David Murphy from Helsinki, Finland / CC BY-SA 2.0)*

TOP TEN BEST-SELLING CONSOLE GAMES OF THE 2010S

The 2010s were marked by the proliferation of smartphones, which took over as the main platform for video gaming. But consoles still enjoyed a healthy patronage, with increasingly sophisticated 'AAA' blockbuster titles benefiting from high development and marketing budgets.

Number ten: *Call of Duty: Black Ops* (2010)
Units sold: 26.2 million
Platform: Multi-platform
Starred big name Hollywood actors Sam Worthington, Ed Harris and Gary Oldman. The player controlled a soldier in the midst of 1960s Cold War intrigue, in a story praised for its dramatic twists.

Call of Duty: Black Ops (2010). *(Treyarch / Activision)*

Number nine: *Call of Duty: Modern Warfare 3* (2011)
Units sold: 26.5 million
Platform: Multi-platform
The *Call of Duty* series of first-person shooter games started in 2003, and – after *Mario* and *Pokémon* – is the best-selling video game franchise of all time. This episode didn't break new ground, but offered a highly polished experience akin to an extended interactive James Bond-style action sequence.

Number eight: *Red Dead Redemption 2* (2018)
Units sold: 26.5 million
Platform: Multi-platform
A huge Wild West action-adventure game. The map covered about 29 square miles, in which you could live out your cowboy fantasies, or unravel the game's epic story.

Red Dead Redemption 2 (2018). *(Rockstar Studios / Rockstar Games)*

Number seven: *Terraria* (2011)
Units sold: 27.0 million
Platform: Multi-platform
The player could explore, build, and fight in a cartoonish two-dimensional world that was randomly generated (or, rather, procedurally generated), creating a different experience every time.

Terraria (2011). *(Re-Logic / 505 Games)*

Number six: *Mario Kart 8* (2014)
Units sold: 27.5 million
Platform: Nintendo Wii U and Switch
A major new instalment of the classic colourful karting series, this time with antigravity driving. These sales figures include the updated version of the game released for the Nintendo Switch console.

Number five: *Diablo III* (2012)
Units sold: 30.0 million
Platform: Multi-platform
A dungeon-crawling role-playing game that was one of the fastest-selling PC games of all time, and has since been ported onto many other platforms.

Diablo III: Reaper of Souls (2014), the first expansion pack for *Diablo III* (2012). *(Blizzard Entertainment)*

Number four: *The Elder Scrolls V: Skyrim* (2011)
Units sold: 30.0 million
Platform: Multi-platform
An open-world fantasy epic set in a huge virtual world populated with rugged mountains, bustling cities, ancient dungeons, and hundreds of characters ranging from warriors and wizards to terrifying dragons.

PlayerUnknown's Battlegrounds (2017). (Whelsko [www.flickr.com/photos/whelsko/]) / CC BY 2.0)

Grand Theft Auto V (2013). (Rockstar North / Rockstar Games)

Number three: *PlayerUnknown's Battlegrounds* (2017)
Units sold: 50.0 million
Platform: Multi-platform
Along with *Fortnite*, this game propelled a new genre into the stratosphere: online 'battle royale', in which dozens of players are dropped onto an island littered with weapons, and the last player alive wins.

Number two: *Grand Theft Auto V* (2013)
Units sold: 115 million
Platform: Multi-platform
One of the most expensive video games ever made, costing $265 million to develop and market. It generated $800 million within 24 hours of its release, making it one of the fastest-selling entertainment products in history (possibly beaten only by the last book in the *Harry Potter* series).

Number one: *Minecraft* (2011)
Units sold: 180 million
Platform: Multi-platform
Minecraft, in stark contrast to *Grand Theft Auto V*, was developed by one Swede, with no budget, and in its first 24 hours (the alpha version) sold exactly 15 copies. The income from the sales of the pre-release version was ploughed back into development, and people kept buying it. A building, exploration and survival game with blocky graphics that captured the imagination of a broad range of gamers worldwide. The best selling non-mobile game of all time.

Minecraft (2011). (Mowenmedia from Pixabay)

TOP TEN MOST PLAYED GAMES OF ALL TIME (UP TO THE END OF 2019)

With the explosion of digital distribution and mobile gaming, a new video game economy emerged, with free-to-play games earning their revenue via ads or optional in-game microtransactions. Which means that the best selling or most profitable games are no longer necessarily the most played games. Some games have been played by a mindblowing number of people, without ever selling a single copy. Incredibly, the top game on this list was played by one in every seven people on the planet.

Most of these figures are based on app downloads and/or registered player accounts, so they are only an approximation of how many people played the game because, for example, one person may have downloaded the game onto multiple devices or created multiple accounts. The last six games on this list all claim half a billion players, making this a top twelve rather than a top ten.

Number seven=: *Tetris* (2012)
Played by: 500 million
Platform: Multi-platform
A new version of *Tetris* for mobile platforms ensured this timeless puzzle game remained one of the most played games.

Number seven=: *QQ Speed* (2012)
Played by: 500 million
Platform: Multi-platform
A Chinese racing game with social elements, starring an anime-style character called Little Mandarin. The British tourist authority VisitBritain collaborated with publishers Tencent to incorporate racing tracks based on locations in England as a way of encouraging Chinese players to visit the country.

Number seven=: *Candy Crush Saga* (2012)
Played by: 500 million
Platform: Multi-platform
A puzzle game with a grid of sweets that the player manipulated to try and make matching rows of three. Publisher King claimed that people have collectively spent over 8 million years playing *Candy Crush Saga*.

Number seven=: *Jetpack Joyride* (2011)
Played by: 500 million
Platform: Mobile
The player navigated a character called Barry Steakfries through an endless corridor of futuristic obstacles by tapping the screen to ignite his jetpack. Steakfries also starred in his own series of musical web cartoons.

Number seven=: *Temple Run* (2011)
Played by: 500 million
Platform: Mobile
The player controlled an explorer running from evil monkeys chasing him across the endless wall of a ruined temple. Many sequels and spin-offs were released.

Temple Run (2011). (Imangi Studios)

Number seven=: *Fruit Ninja* (2010)
Played by: 500 million
Platform: Mobile
Created by Halfbrick Studios, who later made *Jetpack Joyride*. A bunch of fruit got thrown across the screen and the player chopped it in half with a finger swipe.

Jetpack Joyride (2011). (Halfbrick Studios)

Fruit Ninja (2010). (Halfbrick Studios)

Number six: *PUBG Mobile* (2017)
Played by: 555 million
Platform: Mobile
The mobile incarnation of the best selling battle royale game *PlayerUnknown's Battlegrounds* (PUBG). Designer Brendan Greene started off modifying other video games, inspired by the novel trilogy *The Hunger Games*. PlayerUnknown was Greene's online nickname.

PUBG Mobile (2017). *(Sparktour / CC BY-SA 4.0)*

Number five: Dungeon Fighter Online (2005)
Played by: 600 million
Platform: Microsoft Windows
A South Korean side-scrolling fighting game that became internationally popular. The global publisher stopped supporting the game in 2013, but due to popular demand the game was relaunched in 2015.

Number four: *CrossFire* (2007)
Played by: 660 million
Platform: Microsoft Windows
An online first-person shooter hugely popular in South Korea (where it originated) and China. In 2017, the producer of the *Fast & Furious* films announced his intention to make a *CrossFire* film.

Number three: Despicable Me: Minion Rush (2013)
Played by: 900 million
Platform: Mobile
An endless runner similar to *Temple Run*, except the player controlled a minion – a comedic yellow creature from the *Despicable Me* movie and merchandising franchise.

Number two: *Pokémon Go* (2016)
Played by: 1,000 million
Platform: Mobile
Used your phone's camera to superimpose digital creatures into your living room, or wherever else you happened to be playing. From 2018, each month, developers Niantic announced a physical location where an exclusive *Pokémon* monster could be collected, and huge numbers of people gathered in real-life to try and catch one.

Number one: Google Pac-Man Doodle (2010)
Played by: 1,000 million
Platform: Web browser
In 2010, 91 per cent of all web searches were made using Google. For 48 hours in May, Google replaced their logo on their homepage with a playable version of *Pac-Man*, to celebrate *Pac-Man*'s 30th anniversary. Over a billion people played it.

Working on the Google *Pac-Man* Doodle (2010). *(Marcin Wichary / CC BY 2.0)*

CrossFire (2007). *(Smilegate)*

BEST VIDEO GAMES OF ALL TIME

What makes a video game good is, of course, subjective – but there is some critical consensus. Across a selection of 45 lists of the top video games of all time, from magazines and ezines from 1995 to 2019, the following 25 games appeared on at least half of them.

Year	Game	Genre	Publisher	Original platform
1984	*Tetris*	Puzzle	ELORG	Electronika 60
1988	*Super Mario Bros. 3*	Platform	Nintendo	NES
1990	*Super Mario World*	Platform	Nintendo	Super NES
1991	*The Legend of Zelda: A Link to the Past*	Action-adventure	Nintendo	Super NES
1991	*Street Fighter II*	Fighting	Capcom	Arcade
1992	*Super Mario Kart*	Kart racing	Nintendo	Super NES
1993	*Doom*	First-person shooter	id Software	PC
1994	*Final Fantasy VI*	Role-playing	Square	Super NES
1994	*Super Metroid*	Action-adventure	Nintendo	Super NES
1995	*Chrono Trigger*	Role-playing	Square	Super NES
1996	*Super Mario 64*	Platform	Nintendo	Nintendo 64
1997	*Castlevania: Symphony of the Night*	Action-adventure	Konami	PlayStation
1997	*Final Fantasy VII*	Role-playing	Square	PlayStation
1997	*GoldenEye 007*	First-person shooter	Nintendo	Nintendo 64
1998	*Grim Fandango*	Adventure	LucasArts	PC
1998	*The Legend of Zelda: Ocarina of Time*	Action-adventure	Nintendo	Nintendo 64
1998	*Metal Gear Solid*	Stealth	Konami	PlayStation
2000	*Diablo II*	Action role-playing	Blizzard Entertainment	PC
2001	*Silent Hill 2*	Survival horror	Konami	PlayStation 2
2002	*Metroid Prime*	Action-adventure	Nintendo	GameCube
2004	*Half-Life 2*	First-person shooter	Valve	PC
2004	*World of Warcraft*	MMORPG	Blizzard Entertainment	PC
2005	*Resident Evil 4*	Survival horror	Capcom	GameCube
2005	*Shadow of the Colossus*	Action-adventure	Sony Computer Entertainment	PlayStation 2
2007	*BioShock*	First-person shooter	2K Games	PC, Xbox 360

GoldenEye 007 (1997). *(Rare / Nintendo)*

Silent Hill 2 (2001). *(Konami)*

Shadow of the Colossus (2005). *(SCE Japan Studio / Team Ico / Sony Computer Entertainment)*

REFERENCES

A huge thanks to all of the video game journalists, researchers and enthusiasts who have captured so many details about the history of video games with admirable rigour. This book relied on the following sources in particular:

WEBSITES

academic.oup.com
acquired.fm (for the Trip Hawkins quote in Level 4: Companies and Capitalism)
adventureclassicgaming.com
aip.org
allincolorforaquarter.blogspot.com
arcade-museum.com
archive.org (for accessing defunct websites)
aresmagazine.com
arstechnica.com
bloomberg.com
businesstech.co.za
commons.wikimedia.org (for pictures)
complex.com
digitalgamemuseum.org
dorkly.com
eandt.theiet.org
empireonline.com
engadget.com
eurogamer.net
ew.com
fastcompany.com
feministfrequency.com
fhm.com
flickr.com (for pictures)
forbes.com
ft.com
g4tv.com
gamasutra.com
gameinformer.com
games.slashdot.org
gamesdb.launchbox-app.com
gamesindustry.biz
gamesradar.com
gamingbolt.com
goliath.com
guru.bafta.org
hardcoregaming101.net
hawaiibusiness.com
immigrantentrepreneurship.org
independent.co.uk
interestingengineering.com
irishtimes.com
jeuxvideo.com
klov.com
kotaku.com
levelingupyourgame.com
marketwatch.com
masswerk.at
mentalfloss.com
mobygames.com
movingimagesource.us (for the Rochelle Slovin quote in Level 6: Culture and Community)
mtv.com
newyorker.com
npr.org
nypost.com (for the Joe Lieberman quote in Level 6: Culture and Community)
nytimes.com (including for the Katha Pollitt quote in Level 5: Gender and Representation, and the Ronnie Lamm quote in Level 6: Culture and Community)
olympic.org (for the International Olympic Committee quote in Level 6: Culture and Community)
open.lib.umn.edu
pcgamesn.com
pixabay.com (for pictures)
polygon.com (including for the Henk Rogers quote in Level 3: People and Personalities)
pongmuseum.com
popularmechanics.com
pu.nl
ralphbaer.com
replaymag.com
revistagq.com
rockpapershotgun.com
rogerebert.com (for the Roger Ebert quote in Level 6: Culture and Community)
rottentomatoes.com (for aggregated film review scores in Level 6: Culture and Community)
screenrant.com
slantmagazine.com
slate.com
stuff.tv
telegraph.co.uk
thedoteaters.com
thefamouspeople.com
theguardian.com
theverge.com
time.com
uk.ign.com
ukie.org.uk
unite-it.eu
usgamer.net (including for the Tom Kalinske quote in Level 2: Platforms and Technology)
uspto.gov
vgr.com
vice.com
videogamehistorian.wordpress.com
whatculture.com
wikipedia.org
wired.com
yourstory.com
zacfitzwalter.com
zdnet.com

BOOKS AND ACADEMIC PAPERS

Baer, Ralph H., *Videogames: In The Beginning* (Rolenta Press, 2005)

Baran, Stanley, *Introduction to Mass Communication: Media Literacy and Culture* (McGraw-Hill Education, 2014)

Donovan, Tristan, *Replay: The History of Video Games* (Yellow Ant Media Ltd, 2010)

Harris, Blake J., *Console Wars: Sega, Nintendo, and the Battle That Defined a Generation* (Harper Collins US, 2015)

Herman, Leonard, *Phoenix: The Fall & Rise of Videogames Third Edition* (Self-published, 2001)

Kent, Steven, *The Ultimate History of Video Games* (Prima Life, 2001)

Kushner, David, *Masters Of Doom: How Two Guys Created an Empire and Transformed Pop Culture* (Piatkus, 2004)

Newman, Michael Z., *Atari Age: The Emergence of Video Games in America* (MIT Press, 2017) (including for the Toru Iwatani quote in Level 4: Companies and Capitalism)

Sheff, David and Eddy, Andy, *Game Over: How Nintendo Conquered the World* (Cyberactive Media Group Inc/Game pr, 1999)

Williams, Dmitri, *A Brief Social History of Game Play* (University of Illinois at Urbana-Champaign, 2005) (including for the Dmitri Williams quote in Level 6: Culture and Community)

INDEX

Italics indicates there is a picture.

2K Games 113
32X *25*
3D graphics 25, 27, 35, 54-55, 106

Activision Blizzard 72, 113
age ratings see violence
Alcorn, Allan 10, *41*
Apogee Software 54
Apple 42, 59, 71-72
arcade games 9, *10*, *11*, 17, 21, 24, 27, 33, 48-49, *51-52*, 65-67, 74, 83, 95, *100-103*, 113
art 86-88
Atari 15, 23, 25, 40-42, 67-70, 104

Baer, Ralph 11, 12, *39-40*
Bally Midway see Midway
Bandai 27, 67, 72
Bartle, Richard 95
Bertie the Brain 7
Blizzard Entertainment see Activision Blizzard
books see novels
Bushnell, Nolan *40-41*, 42, 67, 69-70

Capcom 21, 113
Carmack, John *54*, 55-57
cartridge 13, 18, 20, *21*, 27, *104-105*
Coleco 13, 39, 63-64, 67
Commodore 25, 70, 107
Condon, Edward 7
consoles
   3DO 25
   3DS 35, 58
   Amiga CD^{32} 25
   Atari 2600 (VCS) 14, *15*, 16, 20, 104
   Atari VCS (2020) 70, *71*
   Binatone TV Master 12
   ColecoVision 14, 64
   Color TV-Game *13*
   Dreamcast 27
   DS 30, *31*, 58, 107-108
   Fairchild Channel F *13*
   Family Computer (Famicom) see Nintendo Entertainment System
   Game & Watch *16*
   Game Boy (and variants) 23, 27-29, 48, 58, 63, *105-106*
   Game Gear 23
   Gamecube 29, 113
   Genesis see Mega Drive
   Genesis Nomad see Sega Nomad
   Home Pong *70*
   Intellivision 14, *15*
   Jaguar 25
   Lynx 23
   Magnavox Odyssey *12*, 39-41
   Magnavox Odyssey^2 (Videopac G7000) *14*
   Master System 20
   Mega Drive (Genesis) *21*, 25, 71, 106
   N-Gage *29*, 30
   NES see Nintendo Entertainment System
   NES Classic Edition 38
   Nintendo 64 24, 27, 106, 113
   Nintendo Classic Mini see NES Classic Edition
   Nintendo Entertainment System (Famicom) 4, *18*, *20*, 24, 48, 104, *105*, 113
   Oculus Rift 35, *36*, 56
   Ouya 31, *33*
   PC Engine see TurboGrafx-16
   PlayStation 25-28, 31, *57-58*, 106, 113
   PlayStation 2 27-31, 58, 105, 113
   PlayStation 3 31, 34, 58, 107
   PlayStation 4 31, 34, *36*, 58
   PlayStation Portable *30*, 58
   Saturn 24-27
   Sega Nomad 27
   SG-1000 18, *19*
   SNES see Super Nintendo Entertainment System
   Super Nintendo Entertainment System 4, *21*, 24-25, 57, 106-107, 113
   Switch *34*, 109
   Telstar 13, *64*
   TurboGrafx-16 (PC Engine) 20-21
   Videopac G7000 see Magnavox Odyssey^2
   Virtual Boy 27
   Wii 30, *31*, 58, *108*
   Wii U 34, *35*, 109
   Xbox *29*, 30
   Xbox 360 30-31, 58, 113
   Xbox One 34, *35*
Consumer Electronics Show 18, 24
crash (market) 14, 17-18, 20, 52, 64, 99, 104
crowdfunding 31, 33, 35, 71, 80

Dabney, Ted *41*, 42, 67
Dungeons & Dragons 47, 54

education 42, 94-95
Electronic Arts (EA) 28, 67, 70-72
Electronic Entertainment Expo (E3) 25, *32*
esports 42, 98-99
Exidy 83

Facebook 35, 46
films 17, 24, 31, 45, 70, 83, 86-90, 112
Firaxis Games 50
floppy disk 4, *16*

Gamergate 81
games
　　Adventures of Lolo, The 74, *75*
　　Alone in the Dark 88-89
　　Angry Birds 89
　　Artillery Simulator 6
　　Assassin's Creed (series) *32*, 76, 89
　　Asteroids 15, *102*
　　Balloon Kid 78
　　Beyond Good & Evil 78
　　Bioshock *32*, 113
　　Black Onyx, The *47*
　　BloodRayne 89
　　Braid 78, *79*
　　Bubble Bobble *67*, 74
　　Call of Duty (series) *30*, *109*
　　Candy Crush Saga *34*, 77, *111*
　　Castlevania (series) *67*, *68*, 113
　　Centipede *101*
　　Chrono Trigger 113
　　Colossal Cave Adventure 44
　　Commander Keen 4, 24, *54*
　　Computer Space 9, *41*, 74, *76*, 92
　　Crash Bandicoot *25*
　　CrossFire *112*
　　Custer's Revenge 83, *84*
　　Dance Dance Revolution 84, 94
　　Dead or Alive 77-78, 89
　　Death Race *83*
　　Defender *102*
　　Depression Quest 81
　　Despicable Me: Minion Rush 112
　　Diablo (series) *109*, 113
　　Doki Doki Panic! 104, *105*
　　Donkey Kong (series) *16*, 44, *52*, 64, *73*, 100, *101*, 102, *103*
　　Doom 4, 24, *55*, 59, 84, 89, 113
　　Dota 2 61
　　Double Dragon 88
　　Drakan: Order of the Flame 78
　　Duck Hunt 105
　　Dungeon Fighter Online 112
　　E.T. the Extra-Terrestrial 70
　　Elder Scrolls V, The see Skyrim
　　Excitebike 104
　　Far Cry 4 76
　　FIFA *38*
　　Final Fantasy (series) 88, 93, *99*, 113
　　Fortnite *98*, 110
　　Frogger *66*
　　Fruit Ninja *111*
　　Galaga *17*
　　Galaxian *101*
　　Galaxy Game 10
　　God of War *28*
　　GoldenEye 007 113, *114*
　　Gotcha *68*
　　Gran Trak 10 *69*
　　Gran Turismo *106*
　　Grand Theft Auto (series) *86*, 107, *110*

Gravity Rush 80
Grim Fandango 113
Guitar Hero 94
Gun Fight 10, *11*
Habitat 95, *96*
Half-Life (series) 46, *59-60*, 113
Halo: Combat Evolved *29*
Head On *66*
Hitman 89
House of the Dead 88
Jetpack Joyride *111*
John Madden Football 71, *72*
Journey (1983) 94
Journey (2012) *37*
Jr. Pac-Man 15
Just Dance 94
King's Quest (series) 23, *24*, 44, *45-46*
Kya: Dark Lineage 78
Legend of Zelda, The (series) 52, *53*, 73, *74*, 88, *104*, 113
Lemmings 4, *107*
Loom *24*
Madden NFL 71
Mafia III 80, *81*
Manic Miner *43*, 44
Mario Bros. 52
Mario Kart (series) *22*, 108-109, 113
Max Payne 89
Mega Man 75
Metal Gear (series) 67, 77, *78*, 113
Metroid (series) *77*, 113
Minecraft *110*
Minesweeper 77
Mirror's Edge 80
Missile Command 15
Mortal Kombat 21, 23, 84, *85*, 88, *100*
Mr Do! 101
Ms. Pac-Man 64, *74*, 102, *103*
MUD 95, *96*
Myst 77
Mystery House *44*
NBA Jam *100*
Need for Speed 89
Never Alone 80, *81*
New Super Mario Bros. (series) *75*, 108
Night Trap 84, *85*
Nintendogs 30, *53*, *107*
Odd Manor 46
Out Run *101*
Pac-Man 15, 17, 52, 66, 74, 88, 102-104, 112
Perfect Dark 75-76
Phantasmagoria 45, *46*
Pitfall! *104*
PlayerUnknown's Battlegrounds *110*, *112*
Pokémon (series) 53, 88-89, *106*, 107, 109
Pokémon Go *37*, 38, 112
Pole Position 88
Pong 10, 12, 40, *41*, 62, 65, 67, 69-70, 92
Pool *8*
Populous 48
Portal (series) 61, 80

Postal 89
Primal 78
Prince of Persia 89
PUBG Mobile see PlayerUnknown's Battlegrounds
Q*Bert 88
QQ Speed 111
Quake 56, 59, 87, 88, 93, 97
Qwak! 68
Radar Scope 51, 52
Ratchet & Clank 89
Rayman 26
Red Baron 49
Red Dead Redemption 2 109
Resident Evil (series) 24, 77, 88, 113
Rhapsody: A Musical Adventure 78
Scribblenauts Unlimited 75-76
Secret of Monkey Island, The 78, 79
Shadow of the Colossus 113, 114
Sid Meier's Civilization (series) 48, 50, 94
Sid Meier's Pirates! 49
Sid Meier's Railroad Tycoon 50
Silent Hill (series) 67, 89, 113, 114
SimCity (series) 4, 48
SingStar 94
Skyrim 109
Snake 27, 28
Softporn Adventure 44, 45
Sonic the Hedgehog 21, 22, 106
Soulcaliber 78
Space Invaders 11, 15, 52, 67, 92, 101, 103
Space Quest 24
Spacewar! 9, 40-41
Star Fox 53
Starcraft 27
Starhorse 101
Street Fighter II 21, 84, 88, 103, 113
Super Columbine Massacre RPG! 84, 86
Super Mario 64 106, 113
Super Mario Bros. (series) 52, 73, 84, 88, 104-106, 113
Super Mario Kart see Mario Kart
Super Mario Land (series) 105-106
Super Mario Maker 2 5
Super Mario Odyssey 80
Super Mario Party 77
Super Mario Run 53, 54
Super Mario World 4, 107, 113
Superbrothers: Sword & Sorcery EP 80
Tank 69
Tekken 89
Temple Run 111, 112
Tennis for Two 9
Terraria 109
Tetris 23, 27, 46-47, 48, 77, 105, 111, 113
The Last of Us 80
Through the Ages 5
Tomb Raider 78, 79, 88
Ultima Online 27, 95
Untitled Goose Game 5
Where in the World is Carmen Sandiego? 80, 94, 95
Wii Fit 108
Wii Play 108
Wii Sports (series) 108
Wing Commander 88
Wipeout 26
Wolfenstein 3D 24, 55, 88
Words With Friends 77
World of Warcraft 89, 95, 97, 113
Worms 6
XCOM: Enemy Unknown 50, 51
Zelda see Legend of Zelda, The

Gamesmaster 88, 89
Garriott, Richard 95
Goldsmith, Thomas T. 6
Google 72, 112
Guinness World Record 31, 59

Halfbrick Studios 111
Hawkins, Trip 70, 71, 72
Higinbotham, William 9
home computer see personal computer

id Software 24, 54-56, 59, 113
internet gaming see online gaming
Iwatani, Toru 74

Kalinske, Tom 25-26
Kassar, Ray 70
Kee Games 69
Kickstarter see crowdfunding
Kinect 31-32, 34
King 111
Konami 66-68, 113
Kōzuki, Kagemasa 67
Kutaragi, Ken 25, 57, 58-59

Lawson, Jerry 14
light gun 12, 68, 105
LucasArts 95, 113

machinima 87
magazines about video games 54, 91-92, 113
Maita, Aki 27
Mann, Estle Ray 6
Mario (character) 21, 48, 51-52, 53, 74, 76, 80, 88, 104-107, 109
massively multiplayer games 95-98
Meier, Sid 48, 49-51
microprocessor 10, 13
Microprose 49-50
Microsoft 18, 27, 30-32, 34, 59, 72, 112
MIDSAC 8
Midway 21, 41, 67, 94
Miyamoto, Shigeru 51, 52-54, 73
mobile phone see phone gaming
Molyneux, Peter 48
movies see films
museums 87
music see sound

Namco 15, 67, 69, 72
NEC 20, 23

Netease 72
Newell, Gabe 59, 60-61
Nimatron 6, 7
Nintendo 4-5, 13, 16, 18-25, 27, 29-30, 34-35, 48, 51-53, 57-58, 62-63, 66-67, 72-73, 88-89, 102, 104-109, 113
Nokia 27, 29
novels 44, 91, 110, 112
Nutting Associates 41, 67

online gaming 27, 56, 95-98, 112
oscilloscope 6, 8, 9

Pajitnov, Alexey 46-47, 48
PC see personal computer
PDP-1 9, 10
personal computers 4, 16-17, 23-24, 33, 47, 54, 59, 113
  Amiga 70, 107
  Apple II 6, 23, 45, 54
  Atari 800 49
  Atari ST 46
  BBC Micro 42
  Coleco Adam 64
  Commodore 64 17, 18, 20, 23, 42
  Electronika 60 47, 113
  IBM PC 4, 23, 113
  MSX 18
  Sol-20 5
  ZX Spectrum 17, 42, 43
  ZX80 42, 43
  ZX81 42, 43
Philips 14, 25
phone gaming 5, 27, 33-34, 37, 48, 53, 109, 111-112
Power Glove 20
Power Pad 20

Quinn, Zoë 81, 82

Race, Steve 26, 58
Rare 75
Ready Player One 89
Robotic Operating Buddy (R.O.B.) 18, 19
Rockstar North 72, 86
Rogers, Henk 46, 47-48
Romero, John 54, 56, 88
Rosen, David 65
Russell, Steve 10

Sarkeesian, Anita 80-81
Sega 18-28, 64-67, 71-72, 106
Sega CD 25
shareware 54-55
Sierra On-Line 4, 23, 44-46, 59
Sinclair, Clive 42, 43-44
smartphone see phone gaming
Smith, Matthew 43-44
Sony 24-26, 28, 30-31, 34, 57-59, 72, 106-107, 113
sound 57, 92-94
Spielberg, Steven 32, 70, 89
Square 67, 113
Star Wars 17
Stealey, Bill 49
Steam 31, 60

Taito 15, 67
Tamagotchi 27, 30
Team17 6
television 6, 11-12, 18, 31, 34, 41-42, 61, 71, 75-76, 78, 83, 88-89
Tencent 72, 111
Tramiel, Jack 70
Tron 89, 90
Trubshaw, Roy 95
TV see television
Twitch 99

Ubisoft 76

Valentine, Don 70, 71
Valve Software 59-61, 113
violence 23, 46, 80, 84-86
virtual reality 27, 35, 57, 61, 99

WarGames 89
Warner Communications 42, 70
Williams, Roberta (and Ken) 44-45, 46
Wizard, The 89, 90
Wozniak, Steve 71
Wreck-It Ralph 89, 90
Wright Will 48

Yamauchi, Hiroshi 48, 62, 63
Yellow Magic Orchestra 92, 94
Yokoi, Gunpei 63